THE FABRIC OF
MIND

A New Understanding of Consciousness and Reality

THE FABRIC OF
MIND

A New Understanding of Consciousness and Reality

ALI SHIRNIA
et al.

Book Cover Image

This book's cover image is a star-filled portrait of the Carina Nebula, 7,600 lightyears away, relatively close in our neighbourhood, a cosmic nursery where new stars are born from massive clouds of dust and gas. Each distant glimmer within this landscape is a star. Our own solar system emerged from a similar star-forming region billions of years ago.

Within such pillars, giant stars lived fast and furious lives, ending in spectacular supernovae explosions. These energetic events forged and scattered the heavy elements that eventually combined to form everything we know, including ourselves. The very atoms that make up our bodies were once part of these stellar giants.

This awe-inspiring image, captured by the James Webb Space Telescope, is not a work of art, but a real image of the heart of a star-forming region in our galaxy.

The image is, in essence, a selfie of our galaxy by our galaxy.

How privileged are we to see it?

Credits:

https://webbtelescope.org/contents/media/images/2022/031/01G7 80WF1VRADDSD5MDNDRKAGY?page=18&filterUUID=91d fa083-c258-4f9f-bef1-8f40c26f4c97NASA, ESA, CSA, STScI

Acknowledgements

My thanks to Helen Tweedale, Hanno Ronte, Marie-Anne Francois and Keyvan Shirnia for your invaluable feedback and steadfast encouragement throughout this challenging process.

Laura Dubey, Phil Mitchell and Bamdad S., your meticulous review of multiple drafts shaped the work's precision. Pat Harris, your initial drive sparked the journey.

Elena Francis, your illustrations make the science feel personal and approachable, as if they are whispering, 'Here, let me explain this part to you.'

My sincere gratitude to you all.

Table Of Contents

Introductions:
The Fabric of Mind

A few years ago, I saw a sketch by the American comedian George Carlin about the relentless forward march of time. As is usual with Carlin, it was a melting pot of wonderful observations, philosophy and some suspect science thrown in:

If he had the choice of how to live his life, he would live it backwards.

He starts out dead ... so there ... death is out of the way and he doesn't need to worry about it for the rest of his life. Next, he wakes up in a nursing home feeling better every day until he is so well that he gets thrown out. He leaves his nursing home only to spend his retirement travelling the world and collecting his pension until he has so much energy he decides to go to work. On his first day at the office his colleagues throw a big party and present him with a gold watch.

He then works for 40 years until he is too young to work. He gets ready for high school; drinking, partying and generally being promiscuous. Then he spends the last nine months of his life floating peacefully in luxury, in a spa-like condition, central heating, room service on tap ... until his life ends in a giant orgasm![1]

It got me thinking, Carlin's whimsical idea might not be as far-fetched as it first seems.

Einstein's special theory of relativity reveals a fascinating fact about time: its flow is not universal. In our universe, the order of events can be subjective, influenced by the observer's motion. What we perceive as the past, present and future becomes a matter of perspective.

Imagine you're standing on Earth, waving. Two astronauts in spaceships travelling at near-light speeds in opposite directions would not agree on when you waved. For the astronaut approaching you, your wave would occur before you consciously registered it. Due to the finite speed of light, her now includes a fraction of your future. Conversely, the astronaut receding from you would see a delayed image of your wave.

This phenomenon, known as the *relativity of simultaneity*, highlights how time is not an absolute but a relative concept. While the cause-and-effect relationship of your decision to wave and the wave itself remains consistent, the precise timing of the wave itself varies for different observers.

At everyday speeds, these discrepancies are negligible. However, as velocities approach the speed of light, the distortions in time perception become significant, demonstrating the malleability of our temporal experience.

From a certain philosophical standpoint, one could argue that all of George Carlin's personal past, present and future exist simultaneously within spacetime, even though he can only experience them sequentially.

The very essence of what constitutes time and reality is puzzling. If the very fabric of time, a dimension we inherently perceive as constant and unwavering, bends and twists depending on our motion, what other seemingly fundamental aspects of reality are relative? Is reality, therefore, just a subjective experience and uniquely tailored for each observer?

This notion challenges our very understanding of existence. If reality isn't fixed, but rather a fluid construct shaped by individual perspectives, then what does this imply about objective reality? Is everything we perceive merely a subjective interpretation of the universe?

These reflective questions lead us to the heart of a fundamental scientific inquiry: the true nature of objective reality.

Suddenly, a short comedy sketch about George Carlin's wish to live life backwards mapped out a series of questions so vast and intriguing that it took me and my collaborators on a journey through maths, physics, chemistry, biology, psychology and philosophy; all the way from the physics of the Big Bang to Alan Turing.

For years, I have had long and meaningful conversations about the nature of reality with Nessie, Iesha and Lilly, our three adopted dogs. They have listened to my existential musings at every morning walk, have been great listeners, critics of ideas ... and such good girls.

My great collaborators, from left to right: Nessie, Iesha and Lilly, returning home from a wet morning walk

Why *The Fabric of Mind*?

For centuries, we believed our universe ticked like a giant clockwork machine. If we knew the precise state of everything at one moment, we could predict the future with absolute certainty; a strictly deterministic order.

Then quantum mechanics entered the picture, and that sense of absolute order started to crack. Why don't we directly experience the probabilistic nature of reality as described by quantum mechanics, the physics of miniscule?

The theory suggests there could be an entire aspect of reality which exists beyond what we can currently see or hear. Could this be right?

Our consciousness filters how we perceive and interpret the world around us. But this subjective lens can distort, omit or embellish the raw data we receive, making it difficult to disentangle reality from personal narrative.

Fields of mathematics, physics, chemistry, biology and humanities each provide essential tools to understand reality. However, their individual capabilities remain limited. Mathematics struggles to model the emergence of intelligence. Physics grapples with the role of the observer in the creation of physical reality and while it offers valuable insights into entropy, it doesn't explain how entropy defines the arrow of time. Biology has yet to fully explain the origins of consciousness.

These outwardly disparate fields, when combined, can offer a comprehensive toolkit for exploring the fundamental fabric of reality.

We explore the physical universe from a unique vantage point: *the evolution of species*. Just as our bodies evolved to survive, our minds developed to understand and navigate the world around us. We'll harness this evolutionary framework, alongside the tools of mathematics, physics, chemistry and biology, to tackle intriguing questions about the nature of physical reality inspired by George Carlin's comedic observations.

Your guide to what is coming up:

Theories of consciousness and theories of mind

The terms *theories of consciousness* and *theories of mind* are often used interchangeably, but they actually refer to two distinct concepts. *Theories of consciousness* aim to understand the nature of consciousness and its origins; the study of the subjective experience of being aware of one's thoughts, sensations, emotions and the perception of the external world. These individual experiences are often termed *qualia*.

Without consciousness, we'd be mere automatons, responding to the world without understanding it. This is our tool for experiencing reality.[2]

Theories of mind, on the other hand, describe the cognitive ability to understand and attribute mental states, such as consciousness, to oneself and other individuals. It involves recognising that others have beliefs, desires, intentions and emotions that may differ from one's own.

Source of minds: exploring the biological basis of consciousness

Based on clinical evidence from patients with brain damage, we can safely infer (health warning: there is no such a thing as safe inference. Inference can suggest strong correlation, but not causation) that consciousness comes from the brain.

Patients with organ transplants don't think differently, but those with brain damage do.

For instance, Nessie underwent keyhole surgery to remove tissues in both her front elbows, and Iesha, due to the advancement of glaucoma, had both eyes removed. However, they both continue to behave the same way. They still enjoy their slices of pears after our walks and become very excited when they hear cheese wrappers.

Additionally, the notion of consciousness emerging from the heart or soul is popular in artistic and cultural literature worldwide. While these theories may be poetic and romantic, they lack predictive power and are inconsistent with observation and other scientific theories. There's no way to actually test them because there is no experiment that could prove them false.

Physicist and mathematician Peter Woit[3] has dubbed such theories 'not even wrong', arguing that a scientific theory must be falsifiable, i.e. it must be possible to devise an experiment that could potentially disprove the concept.

While neuroscientists are making significant strides in understanding the neural basis of consciousness, we focus on exploring consciousness through an evolutionary lens.

The Evolutionary Lens (Chapters 1, 2 and 3)

Over the next three chapters, we develop a unique lens for exploring the nature of reality. This lens suggests that while physics and mathematics offer essential tools for describing the world around us, they are not complete.

We delve into why and how consciousness emerges, contending that it exists not only within humans but also in all animals with a nervous system.

We challenge theories that exclusively place humans at the centre of conscious experiences in the universe, much like how we have erred in the past by positioning humans at the centre of the physical universe.

Instead, we attribute conscious experiences to all organisms with brains.

Nessie, Iesha and Lilly think that *Homo sapiens* have an overdeveloped sense of self-importance.

They all seem delighted by my proposition that the book is about *mind*. They know they are on the spectrum of consciousness and are distant cousins of J. S. Bach.

We then go on to argue that our *perception of reality provides the most reliable description of reality*.

We suggest that our mental representations of the physical world are shaped by the physical world itself, rather than the reverse. In other words, our cognitive frameworks do not determine the fundamental nature of reality but are instead moulded by our interactions with it.

The Evolutionary Lens draws upon the following three profound human scientific endeavours:

1. In 1859, Charles Darwin published a ground-breaking paper titled 'On the Origin of Species by Means of Natural Selection, or the Preservation of Favoured Races in the Struggle for Life'.

   This seminal work outlined Darwin's theory of evolution by natural selection.

2. In 1976, Richard Dawkins[4] introduced the concept of the 'Selfish Gene' in his groundbreaking book of the same name.

   This idea views genes as the fundamental unit of selection, prioritising them over individual organisms or groups.

3. The Integrated Information Theory (IIT) of Consciousness, developed by neuroscientist Giulio Tononi[5] and Christof Koch,[6] provides a scientific framework for understanding consciousness.

    At its core, IIT posits that consciousness arises from the integrated information within a system. According to IIT, the degree of consciousness is directly related to the amount of integrated information.

The Evolutionary Lens holds considerable implications:

*   First, it suggests that consciousness is not a uniquely human trait. Other animals, with a neural network structure, are also conscious.

*   Second, it suggests that consciousness is not a single thing. There are different levels of consciousness, from simple awareness to complex self-awareness.

*   Third, it suggests that consciousness is not fixed. It can change over time, depending on our environment and experiences.

Questions of alien and machine consciousness (Chapters 4, 5, 6 and 7)

Some physicists propose that consciousness actively shapes reality within quantum mechanics, challenging the concept of an objective reality existing independently of the observer. However, the following four chapters will challenge the hypothesis that conscious observers are necessary for quantum measurement to manifest physical reality. Conscious aliens? Conscious machines? Do they create physical reality?

We ask, if consciousness indeed emerges from physical processes, as evidence suggests, could the perceived interaction between observer and observed be an effect, rather than a fundamental cause, within the framework of physical reality?

In Chapter 4, we'll investigate the universe's potential for fostering life and the possibility of how that might lead to consciousness.

Chapter 5 will explore whether consciousness can be reduced to mere computation.

Chapter 6 then tackles the ambitious goal of building conscious machines, critically evaluating the claims and limitations of such endeavours.

Chapter 7 confronts a key question: how might we even recognise consciousness in a machine, if we achieve it?

Questions of reality (Chapter 8)

Our world feels undeniably real; from the objects we touch to the thoughts within our minds.

But is the physical world dependent on our perception of it?

What is the difference between physical and objective reality?

Objective reality has a set of facts that are true regardless of anyone observing or interpreting it. The Earth orbits the Sun, gravity pulls objects towards the centre of the Earth, cells form the basic building blocks of living organisms are all example of objective reality.

Physical reality is the subset of reality that is made up of matter and energy, and the laws that govern them. It's the aspect of the world we can interact with in a tangible way. Rocks, trees, planets, light, sound, energy, chemical reactions are physical reality.

Objective and physical reality often have significant overlap. For example, we might believe the existence of the Sun is an objective fact because its physical reality can be observed and measured.

Modern physics, with theories such as quantum mechanics and general relativity, has begun to challenge traditional notions of objective reality. In the quantum world, the act of observation seems to affect physical systems, and the nature of time and space becomes less clear-cut.

Given that our understanding of physical reality is shaped by our observation and measurement tools, to what extent can we truly claim an objective grasp of it?

Is our world physical? (Chapter 9)

In quantum physics, the role of the conscious observer stands as a controversial concept, challenging our fundamental understanding of reality. At its heart lies the phenomenon known as the observer effect, wherein the act of measurement or observation fundamentally alters the behaviour of quantum particles.

When an observer uses an apparatus to observe these particles, they transition from a state of potentiality, existing in a range of probabilities, to a definite state, a process described as wave function collapse by the father of quantum mechanics, Niels Bohr.

Why does our micro-world depend on a conscious observer for its existence?

This intriguing manifestation raises deep questions about the nature of consciousness and its relationship to the fabric of the universe.

Does consciousness play a pivotal role in shaping our reality? Do our observations actively participate in the creation of the world around us? This notion blurs the line between the subjective world of human consciousness and the physical reality of the quantum world.

Is time real? (Chapters 11 and 12)

Why do equations of motion in physics allow time to run backwards or forwards, while time in our brains only moves forward? Why is there an arrow of time in our heads? Yesterday is gone and will never return; the present moment is fleeting, but what does *now* even mean? And tomorrow, well, we must wait for that.

The present continuously slips into the past, while the future becomes the present. And so, the cycle continues.

What makes a good theory?

A good idea should be as simple as possible while still explaining the phenomena it addresses.

In short, a good theory is one that is consistent with evidence, predictive, testable and fruitful. It is a powerful tool for understanding the world around us, and it has helped us make great progress in science.

For example, the belief known as Young Earth creationism says that Earth was created 6,000 years ago, a notion that conflicts with scientific evidence.

Then we ask, 'What about dinosaur fossils which have been carbon dated accurately and are hundreds of millions of years old?' Answer: 'Oh, well, it was created and put in place at the time of the creation of Earth.'

'Did oxygen not appear 2.45 billion years ago, during a period of time known as the Great Oxidation Event?' Answer: 'Oh, it was created and put in place at the time of the creation of Earth.'

'How about the oldest rock ever found on Earth, the Acasta Gneiss, which is located in the Northwest Territories of Canada and is about 4.031 billion years old?' Answer: 'Oh, well, that was all created 6,000 years ago and put in place at the time of the creation of Earth.'

Unlike a good theory, Young Earth creationism relies on unprovable assumptions and ignores contradicting evidence. Instead of offering explanations, it avoids them with circular reasoning.

Lilly is concerned. 'Have we just lumped most theories of consciousness together with Young Earth creationist and Flat Earth theories?' I explain to her, 'They exhibit the same traits so they are in the same genre.'

A concordance check: how would we know we are on the right track?

In sciences, such as particle physics, cosmology or palaeontology, where we have no way of setting up experiments to prove a theory true or false, a system of truths (concordance model) is used as a model that is consistent with a wide range of theoretical and observation results even if some pieces of the puzzle are still missing.

For instance, in particle physics, the standard model is a concordance model that is consistent with a wide range of observations from particle accelerators and astrophysical observations.

In philosophy, the concordance model is used to argue for the truth of a theory or belief by showing that it is consistent with other well-established theories and beliefs. This is done by finding concordances, or points of agreement, between the different theories and beliefs.

At the end of each chapter, we will test our proposition based on the concordance model.

My collaborators cannot wait any longer and they let me know, very loudly, we need to move on. However, before we do, it's important to note that we often omit definite articles like 'the' before terms such as 'brain', 'reality' or 'mind' when the nature of the concept itself is still under investigation.

Let's start building The Evolutionary Lens.

Chapter 1:
The Evolutionary Lens

We suggest that consciousness, having evolved as an adaptation, provides organisms with a unique perspective to comprehend the underlying essence of reality.

Nessie, Iesha and Lilly are concerned. 'We don't even know what consciousness is, so how can it possibly tell us something about the nature of our world?'

I explain, 'We treat consciousness as a black box here.'

Think of a black box as simply a tool or a system, the internal workings of which are unknown to us, yet we can still learn from its observable outputs. In this way, consciousness is analogous to the James Webb Space Telescope (JWST).

While we can utilise JWST as a powerful tool to explore the composition of the universe even without fully understanding its inner workings, we acknowledge that deeper knowledge of its capabilities, such as its specialisation in infrared wavelengths, would enhance our ability to detect faint, distant objects.

Similarly, we acknowledge the limitations of using consciousness as a 'black box', yet we argue that it remains our most valuable tool for deciphering reality.

This does not negate the importance of investigating the nature of consciousness itself. In fact, such research could unlock even deeper insights into our understanding of the universe.

Consciousness as a tool

The Evolutionary Lens says that conscious behaviour does not exist independently of a brain. It is not a *ghost in the machine*, a *soul* or an *unfathomable force* in the universe. Rather, it is the result of integration of cognitive processes such as sensory perception, attention, memory and decision-making within brain.

Consciousness didn't pop out of nowhere. It's a product of how brain has developed to understand physical reality, keep track of itself and make predictions for the future.

Think of it as the organism's ultimate decision-making tool, using everything it experiences to optimise survival. The organism deploys its senses and problem-solving skills to perceive threats, predict and understand the behaviour of others, detect deception and manipulation. By attributing consciousness to both allies and adversaries, the organism gains insights into their behaviour, fostering trust and cooperation in social interactions.

What is *The Evolutionary Lens*?

For approximately 500 million years, since the first appearance of animals with integrated neural networks, consciousness has been tuned and tested to destruction – literally.

Each brain is a device, with built-in input and output data and control channels. Its responsibility is to preserve genetic information through whatever mechanism it finds appropriate;

locate friends, foes, mates, food and shelter, directed by the goals of survival and replication.

Each brain is individually trained through sensory input, repetition, reinforcement, education, experience, etc.

There are an estimated 8.7 million species [1] with approximately 20 quintillion,[2] [3] [4] or $2x10^{19}$, individual animals who inhabit planet Earth today. That is $2x10^{19}$ brains of all sorts of complexities, from the simplest organisms to the most advanced mammals.

We argue that all inhabitants of Earth, all these countless brains and organisms share a common physical reality. Adapting to environmental changes requires them to perceive the same external stimuli, interpret those cues constantly and adjust their behaviour accordingly.

Successful navigation of a shared environment demands that organisms perceive largely the same external stimuli. Failure to do so consistently would lead to extinction, not the evolutionary flourishing we observe.

This simultaneous adaptation demonstrates a unified and objectively experienced physical world.

The physical world is the ultimate proving ground for all animal brains. It is highly improbable that 20 quintillion different realities are, somehow, all coordinated to form the tree of animal life.

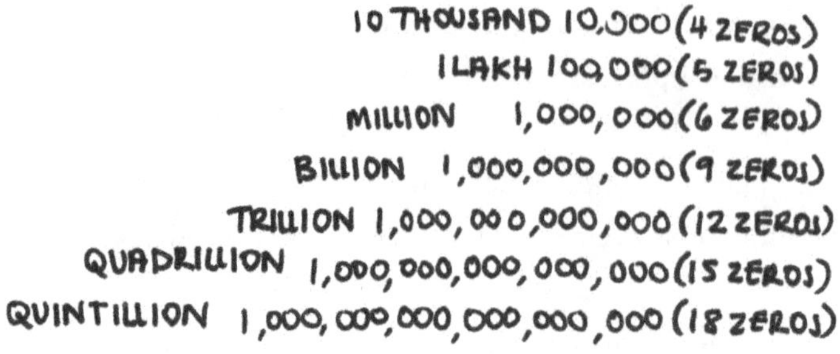

10 THOUSAND 10,000 (4 ZEROS)
1 LAKH 100,000 (5 ZEROS)
MILLION 1,000,000 (6 ZEROS)
BILLION 1,000,000,000 (9 ZEROS)
TRILLION 1,000,000,000,000 (12 ZEROS)
QUADRILLION 1,000,000,000,000,000 (15 ZEROS)
QUINTILLION 1,000,000,000,000,000,000 (18 ZEROS)

Figure 1: 20 quintillion is a big number

We conclude there is a single physical world which serves as the training data for brains of all inhabitants of planet Earth, each a self-determining, independent, internally self-regulating and autonomous system.

By experiencing our world through our senses, we're interacting with the physical reality that has shaped our brains.

A concordance check: is brain a glitchy device?

The concept that consciousness constructs its own personal reality fails to account for the overwhelming evidence that points to a shared, physical world.

Would a chase in the park be as thrilling if every dog and squirrel perceived a unique reality, independent of the others?

The rigorous evolutionary training, designed to ensure survival and replication, demands accurate perception and interpretation of external stimuli. To dismiss the existence of this physical reality would imply an extraordinary and implausible coordination between countless independent, self-determining systems.

The concordance model in cosmology reinforces this perspective.

It paints a picture of a universe governed by fundamental forces, laws of physics and vast amounts of matter and energy. This model provides an additional framework for understanding the consistent, ordered and rule-bound nature of the reality that we experience.

While our understanding of both consciousness and the universe remains incomplete, the evidence converges towards a physical world that exists independently of our perception of it.

In a nutshell

The evolutionary story of consciousness reveals it to be a remarkable instrument, uniquely tuned to the true nature of the world around us. Brain is the product of reality and thus the most reliable tool we have to understand it.

What if evolution has caused all brains to develop the same delusion of an external reality?

We think it's far more parsimonious to believe an external world exists than to imagine the insane level of coordination needed for such a complex self-deception across countless species.

Chapter 2:
Science Of Consciousness

Is there such a thing as science of consciousness?

Your brain is constantly predicting the future. Not in a psychic way, but milliseconds ahead, filling in gaps in your perception to create a seamless experience. It does this to help you survive. Imagine trying to cross the street if the world moved in jerky stop-motion!

But are these predictions merely a survival tool, or do they form the very foundation of our perceived reality?

For centuries, philosophers deliberated the question of consciousness. Was it a gift from a higher power, or simply an illusion created by brain? Today, theoretical and experimental scientists are taking over the investigation.

In Chapter 1 we positioned consciousness as a finely tuned tool for exploring our universe's core principles. This chapter provides a brief context and a concise overview of consciousness theories, tracking their evolution from early concepts to the forefront of research. We delve into a promising scientific framework that underpins our exploration, a framework that positions consciousness not merely as a biological quirk, but as a powerful survival tool shaped by evolution.

We examine how brain's ability to integrate information from different sensory mechanisms, memory and cognitive processes might hold the key to understanding our subjective experience.

Consciousness is a sophisticated instrument honed by evolution, enabling us to perceive and interpret the fundamental nature of reality.

A precise definition of consciousness

Throughout history, countless philosophers, scientists and thinkers have grappled with the definition of consciousness, yet a singular, universally accepted understanding remains elusive.

Susan Blackmore,[1] a leading psychologist and author in the field, suggests that consciousness is not a single thing, but rather a collection of different abilities such as:

- *Sensation:* Our capacity to perceive the world through our senses.

- *Attention:* Focusing on a specific aspect of our sensory experience.

- *Memory:* The power to store and recall information.

- *Thought:* Processing information, reasoning and making decisions.

- *Self-awareness:* Recognising ourselves as distinct, unique individuals.

Blackmore says that consciousness is an emergent property of brain, meaning that it arises from the interactions of many different brain processes.

She rejects the notion that it is a separate substance or entity and argues against the need to invoke any supernatural or mystical forces to explain it.

Blackmore's definition is widely acknowledged as the most accepted in the field. It is simple, straightforward, and captures the essential features of consciousness without becoming overly complex or mired in jargon.

The intersection of consciousness and reality

The challenge of comprehending consciousness, as we attempt to unravel the mechanisms underlying our experiences, is eloquently encapsulated by Schrödinger:

> "Strange fact that on the one hand all our knowledge about the world around us, both that gained in everyday life and that revealed by the most carefully planned and painstaking laboratory experiments rest entirely on immediate sense perception.

> While on the other hand this knowledge fails to reveal the relations of the sense perceptions to the outside world so that in the picture, or model, we form of the outside world, guided by our scientific discoveries all sensual qualities are absent."

E. Schrödinger, *Mind and Matter 1958*

Nessie, Iesha and Lilly stop and stare at me, 'Isn't he the guy with the lovely cat? How is she?' I reply, 'She is both fine, and not. I don't dare look.'

Some argue that solely focusing on brain's internal workings misses a crucial element of consciousness.

Michio Kaku, an American theoretical physicist, suggests that this realisation prompts some scientists to reconsider the methods employed in studying brain, resulting in a proliferation of unconventional and imaginative theories attempting to explain consciousness.

He makes a valid point. The spectrum of theories on consciousness spans from the *supernatural* to the *fundamental* and *incomprehensible*. It ranges from the idea that the universe itself is conscious, to the belief that only Zeus possesses consciousness, to the notion that consciousness defines the universe, or that the universe is finely tuned for consciousness by a supernatural being.

Some even propose that to possess a mind, one requires something beyond a brain, while others argue that consciousness is not real; it is merely an illusion.

The zeitgeist of theories of consciousness

When we refer to *zeitgeist*, we mean the spirit of the age, or the prevailing intellectual and cultural climate of our time. This encompasses the set of assumptions and beliefs that dominate a particular era.

Such prevailing notions can significantly influence the perspectives of scientists and philosophers and the theories they develop.

In the 19th century, the zeitgeist was shaped by the rise of scientific materialism, which posited that all phenomena, including consciousness, could be explained in terms of physical processes.

This paradigm shift led to the development of several *physicalist theories*, including the proposition that consciousness arises as a product of brain's electrical activity.

In the 20th century, the zeitgeist was shaped by the emergence of cognitive science, which influenced the development of numerous cognitive theories. Among these theories is the proposition that consciousness arises from brain's capacity to represent and process information.

The zeitgeist is continually evolving, leading to shifts in the dominant theories of each era. In the present day, the zeitgeist is shaped by the ascendancy of artificial intelligence and a growing understanding of brain systems.

These factors have contributed to the emergence of new theories that highlight the significance of attention, the self and environmental factors in shaping consciousness.

The contributions of the great Greek philosophers to the field remain relevant today. The questions they raised laid the foundation for much of the modern debate, which continues to be explored by philosophers and scientists alike.

The beginning of the chain of studies is credited to Thales (624–546 BC), who wrote about the nature of the universe and our perception of reality. Pythagoras (570–495 BC), Plato (428–348 BC) and Aristotle (384–322 BC) all made significant contributions to the study of mind and body.

Plato believed that mind is a distinct entity from the body and considered it to be immortal. He thought that consciousness arises from the soul, which he viewed as a rational and eternal essence.

Plato explored the nature of reality and knowledge in his renowned Allegory of the Cave, presented in Book VII of his work *The Republic*. This allegory depicts two separate realms: one that is observable and another that represents true existence. It has been interpreted as a metaphor for human perception and understanding.

Aristotle held the view that mind is not separate from the body but rather a function of it. He developed theories on perception, memory and reasoning, which laid the groundwork for the study of cognitive processes. Aristotle argued that consciousness is simply the ability to process information and make decisions.

In their writings, Pythagoras, Plato and Socrates explored the concept of free will, arguing that individuals could only make truly free choices if they were conscious of their own thoughts and desires. They debated whether this capacity for self-awareness was unique to humans or if it extended to all living beings.

Nessie looks at me in amazement at the works of these giants of philosophy and science which are debated to this day.

I explain to her, 'This is why they are called Great Greek Philosophers. This is what greatness looks like.'

In Persia, Omar Khayyam's *Rubaiyat*, first published in the 16th century, quickly gained popularity throughout the Islamic world. Characterised by their wit, their humour and their insights into the human condition and mind, the Rubaiyat continues to captivate readers to this day.

René Descartes (1596–1650) proposed the first recorded theory of consciousness, known as Dualism. Descartes' theory was based on his famous cogito ergo sum argument, meaning I think, therefore I am. He argued that the ability to contemplate one's own existence demonstrated the existence of a non-physical substance, which he referred to as consciousness or mind.

Over 2,500 years of contemplation, numerous threads have emerged in the quest to understand the nature of consciousness, each with its own strengths and weaknesses.

Some theories emphasise the physical aspects of consciousness, focusing on brain's role in processing information. Others delve into the mental aspects, exploring our thoughts, feelings and experiences.

The body of work on consciousness is vast, with a wealth of literature available for those interested in exploring this subject (see References). Given the breadth of research, it is worthwhile to categorise the contributions of the leading contenders in the field today.

Evolutionary theory of consciousness: While Charles Darwin never fully developed a theory of consciousness, he did write essays on the topic and engaged with the problem of explaining the phenomena within a purely materialistic framework.

In his book *The Descent of Man* (1871), Darwin[2] wrote:

> "The mental powers of very low animals, such as the Condylactis, are probably of a very humble order of nature. But in all the higher animals, especially in mammals, these powers are of the highest importance, as they relate to the welfare and very existence of the individual.

> No animal can survive during a severe winter, unless it possesses warm clothing; and no animal can procure food unless it can discover and secure it. Both these powers, as well as many others, are directly connected with the intellectual faculties."

In discussing the role of consciousness in evolution, Darwin proposed that it provided animals with a distinct advantage in the struggle for survival. He argued that animals capable of anticipating danger or learning from their experiences were more likely to survive and reproduce.

(*The Descent of Man*, 1871):

> "At what age does the new-born infant possess the power of abstraction, or become self-conscious and reflect on its own existence? We cannot answer; nor can we answer in regard to the ascending organic scale ... The half-art, half-instinct of language suggests evolution shapes language alongside instinct."

In his autobiography, 1887:[3]

> "Consciousness is the most important of all the great facts of life. For evolutionary continuity, there is nothing unique about nature; there is no difference between creatures on the tree of life and sentience is a continuum."

Richard Dawkins,[4] a British evolutionary biologist, has extensively written about the subject of consciousness. He advocates the view that consciousness is an emergent property of brain, rejecting the notion that it is a separate substance or entity.

In his book *The Selfish Gene* (1976), Dawkins argues that consciousness is a product of natural selection. He says that it emerged through the evolutionary process because it provided our ancestors with advantages in survival and reproduction.

Awareness of danger allowed them to avoid predators, while the ability to plan facilitated finding food and mates.

According to Dawkins[5] consciousness is not necessary for life; rather, he views it as a complex adaptation that evolved in some animals, not as an essential component of life itself.

In *The God Delusion*, he says:

> "Consciousness is a spectrum, and there is no clear line between where it begins and where it ends. Even within humans, there is a wide range of consciousness, from the deep sleep of a coma patient to the intense awareness of a meditative master."

Recent neuroscience research confirms the existence of a spectrum of consciousness. We all share the same brain architecture, information flow, integration and motivation for survival and replication.

Nessie, Iesha, Lilly and I are all on the spectrum.

As we will discuss later in this chapter, recent advances in IIT provide support for the spectrum of consciousness we observe in ourselves and other animals. This research focuses on the brain's capacity to integrate information as a key factor in conscious experience.

Dualism is the theory proposing that consciousness is a separate substance or entity from the physical body. René Descartes first proposed this theory in the 17th century, and it has been defended by many philosophers since then.

Descartes argued that the mind and the body are two distinct substances that interact with each other in some way.

One of the leading contributors to the field today is David Chalmers,[6] [7] [8] NYU professor, who proposes that there are two different kinds of consciousness: phenomenal consciousness and access consciousness.

Phenomenal consciousness refers to the subjective experience of being aware of something, such as the feeling of what it is like to see red, hear a dog bark, or feel pain. Chalmers argues that phenomenal consciousness is a fundamental property of the universe and cannot be reduced to physical properties.

Access consciousness refers to the ability to be aware of something and to utilise that information to guide our thoughts and actions, including the ability to reflect on our experiences and communicate them to others. Chalmers says that this property is a product of brain and can be explained in terms of physical properties.

He goes on to assert that phenomenal consciousness constitutes the *hard problem* of consciousness, which involves explaining how physical processes in brain give rise to subjective experiences.

Chalmers argues that the hard problem cannot be solved by *physicalist* theories alone. According to him, consciousness simply exists, and we must accept it for what it is — an irreducible and fundamental force of nature.

The Evolutionary Lens takes a different approach to Chalmers' *hard problem*. It argues that consciousness evolved precisely to solve the problem of decoding the complex physical reality that shaped us.

In the immediate aftermath of the Big Bang, the universe was an incredibly hot and dense state. Energy and fundamental forces were initially unified. As the universe expanded and cooled over fractions of a second, those forces began to separate and take their current form. This separation allowed particles like quarks and electrons to form, the building blocks of the matter we know today.

Fundamental forces include electromagnetism, the strong nuclear force, the weak nuclear force and gravity. They are considered fundamental, meaning they cannot be further reduced to other physical phenomena.

If consciousness is fundamental, then Chalmers must explain how it ever emerged from pure energy at the dawn of our universe. How did the Big Bang lead to the *ghost in the machine*?

We are all puzzled by Chalmers' proposition. Four tilting heads attest to it.

Dualism assumes too much and explains too little. It is not even wrong, because it cannot be falsified.

It is unanimous; we believe that dualism is a defeatist point of view in philosophy and represents a barrier to the study of consciousness.

The perspective that either consciousness is a fundamental force of the universe or that we will never understand it because it does not emerge from the realm of physics has impeded progress in the field for many years.

Fortunately, recent laboratory studies of brain have revitalised the field, leading to significant progress in our understanding.

In response to such criticism, Chalmers proposes that *panpsychism* is a possible answer to his postulate that consciousness is fundamental.

Panpsychism is the theory proposing that consciousness is a fundamental property of the universe, existing within all entities, including inanimate objects.

This concept was initially introduced by Leibniz in the 17th century and has since been supported by various philosophers and scientists. Leibniz said that consciousness permeates all matter, albeit more prominently in certain entities than others.

Panpsychism addresses the question of how consciousness arises from pure energy during the moment of the Big Bang. It suggests that every particle in the universe, including quarks, electrons and photons, possesses some form of consciousness, contributing to a universal awareness.

While this perspective is poetic and intriguing, it remains unverifiable.

Proponents of panpsychism do not see the universe as solely a vast quantum computer. Instead, they attribute consciousness to every particle, beyond merely assigning quantum information to them as Richard Feynman, Nobel Laureate and brilliant quirky physics icon, did.

By a unanimous show of paws and hand, we've reached a consensus: dualism and panpsychism are not robust theories. They lack descriptive power and fail to address existential questions adequately.

Moreover, they are not testable and offer limited utility in predicting new phenomena, rendering them unsuitable for scientific research.

Naturalism is the theory proposing that consciousness arises solely from the physical processes of brain. This concept was first articulated by Thomas Hobbes in the 17th century and has since garnered support from numerous scientists and philosophers. Hobbes argued that consciousness is intricately tied to the complex physical workings of brain.

John Searle[9] [10] [11] is a prominent philosopher and professor of mind. He says consciousness is best understood as a biological phenomenon emergent from the operations of brain. He rejects the idea that consciousness can be reduced to purely computational or functional processes, instead emphasising the importance of brain's biological structure and processes in generating conscious experience.

Searle's approach highlights the intrinsic connection between consciousness and the physical workings of brain, distinguishing his views from purely computational or abstract models of mind.

John Searle's *naturalism* view raises questions and offers limited explanations. Why does he presume that consciousness exclusively emerges from biology? Could consciousness arise in biological forms beyond Earth, elsewhere in the universe? What makes Earth's biology uniquely suited for the emergence of consciousness? And could consciousness manifest if we were to cultivate neurons in a laboratory setting?

Naturalism is a really big leap in faith and difficult to prove.

Functionalism is the theory that consciousness is a functional property of brain. This theory argues that it is not a substance or entity, but rather a set of functions that brain performs. For example, consciousness might be the function of brain that allows us to be aware of our surroundings, to make decisions and to experience emotions.

Daniel Dennett,[12] [13] [14] Professor of Philosophy at Tufts University, was a prominent philosopher and cognitive scientist who is known for his work in the field. He argued that there are complex phenomena that emerge from the interaction of many different factors, including brain, the body and the environment.

Dennett's *multiple drafts* model is a theory of how consciousness emerges from brain. The model argues that it is not a single thing, but rather a collection of different drafts or versions of reality that are constantly being generated by brain.

He said, "Consciousness is not solely about knowing the truth, but rather about enabling effective action in the world. Organisms are continually striving to maximise their fitness, and consciousness serves as a mechanism to assist them in making better decisions towards this end."

My fur collaborators seem confused by Dennett's argument. They think, 'Can an organism become fitter, make better decisions and be more effective spending energy and time selecting a *draft of reality?* Afterall, this is what Dennett is implying.

Lilly thinks, 'Why have many drafts of reality when one would do? How do squirrels select a draft amongst many? Does that not make all other drafts useless?'

She is right. In a desperate struggle, sifting through countless drafts to find the right one could spell doom.

I explain, 'There is no evidence to suggest the claim that an organism can instinctively and accurately choose a single, true representation of reality from an endless number of incorrect ones.'

In attempting to reject the traditional idea of a centralised *Cartesian theatre* in the brain where all conscious experience comes together for a unified *show*, Dennett does indeed introduce a paradox when considering the evolutionary advantage of deceptive mechanisms. In his theory, consciousness is not necessarily about knowing the ultimate truth but rather about facilitating effective action.

Consciousness, as a tool for survival, likely evolved to provide an accurate, rather than false, representation of the world.

Isn't the fundamental purpose of possessing a conscious brain to excel in our world? How would an organism survive the flow of constant misinformation? There is only one *truth* and infinite number of *untruths*!

Such philosophical explorations provide valuable questions about the nature of consciousness, but they lack quantifiable ways to understand how they manifest physically.

By the way, what does *truth* mean? We only use it here because Dennett refers to physical reality as *truth*. In our four heads we replace it with physical reality.

We can only digest *truth* in formal logic.

According to The Evolutionary Lens, our perception filters physical reality, presenting us with a subset relevant to our survival; a useful version of physical reality, not the entirety of it.

To move beyond abstract concepts and tackle consciousness in a scientific manner, we need a way to measure and quantify it.

Towards science of consciousness: information theories of consciousness

While philosophy offers tantalising questions, The Evolutionary Lens demands a way to measure and quantify consciousness. Enter IIT, a promising framework that focuses on brain's capacity to process and integrate information to potentially explain this phenomenon.

We need to create a system which accurately gauges the level of consciousness in organisms without overestimating or underestimating it. If we treat all organisms as if they were clockwork, we would miss everything that is important and exciting about them. If on the other hand we treat a clock as if it is a complex conscious object, we would waste years of valuable research time on gobbledygook.

William James,[15] the father of American psychology, claims that consciousness requires some degree of goal-directedness, some ability to take different paths to get to the goal. The organism must also be capable of taking actions not completely determined by local circumstances. This ability dictates where the organism is on the spectrum of consciousness.

In recent years, there has been growing interest in information-processing theories of mind. These theories propose that awareness is a product of brain's ability to process and integrate information.

What is consciousness?

IIT, developed by Giulio Tononi, University of Wisconsin–Madison and Christoph Koch, Allen Institute for Brain

Sciences,[16] [17] [18] [19] [20] suggests that the more information that is integrated, the more conscious the system is.

According to ITT, conscious states are highly differentiated and information-rich and are highly integrated as a whole; the details are all bound together as a unified phenomenon.

In short, according to Christoph Koch, "Consciousness is what goes away when we fall into a dreamless sleep! What goes away, seems to be, literally, everything. We don't see anything, we do not hear anything, we do not feel anything, we have no emotions and no thoughts. As far as we are concerned, there is nothing at all. When we regain consciousness, by either waking up or dreaming, we experience consciousness; sounds, images, thoughts and feelings."

Any organism with a neural network system can experience consciousness, and the degree of integration of systems, cognitive, sensory, memory, etc. define the spectrum of consciousness.

Information integration is calculated by measuring the amount of information that is shared between different parts of the system; the various parts of a brain which collaborate to create a unified conscious experience.

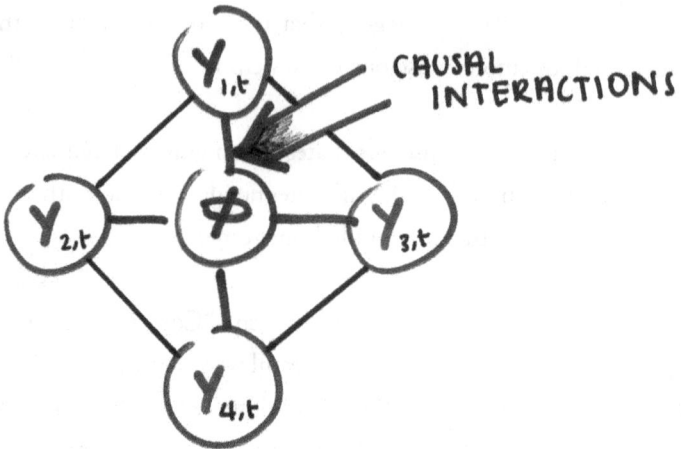

Figure 2: Neural networks causal structure

Tononi and Max Tegmark,[21] [22] [23] Massachusetts Institute of Technology, show that in this theory, *Qualy*, Φ*(phi)*, represents the integrated state of the system at a particular time. They tackle the complex matter of quantifying consciousness levels by introducing a method that calculates it within a system.

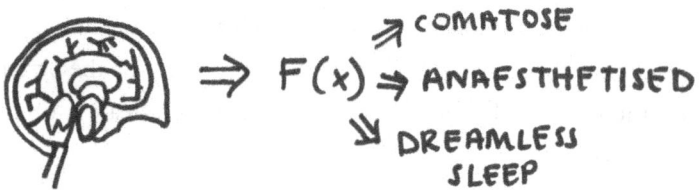

Figure 3: Levels of consciousness

How is Φ calculated? Each y is an integrated unit of the brain which contributes to consciousness. All units of y together sum to the value of Φ.

If the causal link is cut, change in the value of Φ denotes the contribution of specific y to the total system.

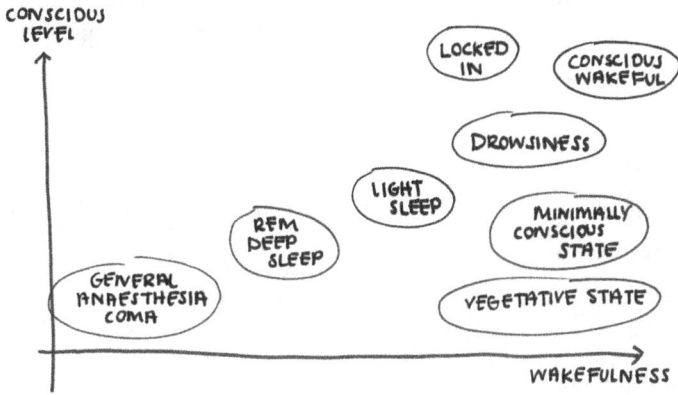

Figure 4: Levels of consciousness

Figure 4 shows Φ has a different value for comatose, anesthetised and dreamless sleep. The value denotes the spectrum of consciousness in any organism.

Consciousness as the 'Ultimate Decoder'

IIT's focus on integration explains why brains, and not clocks or rocks, are the locus of consciousness.

As we walk in the park, enjoy the view of the lake, the trees, as doggie friends come to say hello and ask for treats, we are all in a responsive state to inputs from our sensory apparatus, and we register that information and integrate it all.

We are in a definite qualitive state when we form those integrations. IIT relates consciousness to the number of elements in this complex structure.

The constellation of all systems in our brains together constitutes the spectrum of *qualia* and consciousness. Therefore, there are boundaries for each one of us, where due to construction and integration of our individual brain systems, we can only move in our qualia spectrum.

So, if Nessie, Iesha, Lilly and I attended a neuroscience seminar at Cambridge Medical School, our integrated systems would be overwhelmed. There is a lot that we would not be able to comprehend, synthesise, say or produce. To each one of us, our appreciation of every concept at the seminar is limited by the state of our integrated brain systems mechanism.

In 1998, Christoph Koch and David Chalmers made a bet on whether science would have an explanation for consciousness by 2023. Koch wagered a case of wine that it would, while Chalmers bet that it would not.

Koch was optimistic that the next 25 years would see significant progress in our understanding of consciousness. He believed that scientists would be able to identify the neural correlates of consciousness, or the patterns of brain activity that underlie conscious experience.

The bet was settled on 23 June 2023, at the annual meeting of the Association for the Scientific Study of Consciousness (ASSC) in New York. Both Koch and Chalmers agreed that science had not yet achieved a clear understanding of consciousness. Chalmers was therefore declared the winner of the bet.

Koch's setback in this particular experiment doesn't negate the significance of IIT's concept of quantifying consciousness. The immense computational challenge in calculating Φ for complex systems highlights the intricate nature of consciousness itself, rather than undermining IIT's validity. This result doesn't necessarily validate Chalmer's position either, as his hypothesis remains outside the realm of empirical testing. Chalmers argues that consciousness is a fundamental property of the universe. He does not explain how consciousness emerges as a fundamental property of the Big Bang.

We now know that by using novel techniques or platforms such as quantum computers, we can logarithmically accelerate time to solution for some intractable calculations. Recent work on visualisation of the inner working of neural networks paves a path for a universal quantum computer algorithm called Quantum Fourier Transform (QFT) to calculate the value of Φ in a reasonable, i.e. useful time.

The debate is likely to continue for many years to come. However, the wager between Koch and Chalmers has helped to raise awareness of the topic. The endeavour to identify all the components of brain that contribute to the integration of information goes on.

I tell my collaborators, 'I think it is too early to judge Koch as the loser. Efforts to map the brain using IIT is moving really quickly.'

Rendering by Philipp Schlegel (*Drosophila* Connectomics Group, Cambridge)

Figure 5: The first neuronal wiring diagram of
a whole adult brain, a female fruit fly

In Figure 5 connections between neurons of an adult fruit fly are mapped by acquiring and analysing electron microscopic (EM) brain images.

Here, Dorkenwald et al. and Schlegel et al., FlyWire Consortium,[24] present the first neuronal wiring diagram of a whole adult brain, containing 5×107 chemical synapses between ~130,000 neurons reconstructed from a female *Drosophila melanogaster*, fruit fly.

They demonstrate the integration of synaptic pathways and the analysis of information flow from inputs (sensory and ascending neurons) to outputs (motor, endocrine and descending neurons), across both hemispheres, and between the central brain and the optic lobes.

The best part?

The technologies and open ecosystem of the FlyWire Consortium has set the stage for future large-scale connectome projects in other species.

We had a bit of vertigo when we studied IIT. My collaborators are delighted that at last we have a scientific approach to attributing consciousness to all organisms with neural networks.

We take advantage of the theory's excellent idea, i.e. measurable integration of information as a scale for consciousness. The Evolutionary Lens suggests consciousness evolved to solve complex problems. IIT offers a means to *measure* the complexity that brain evolved to tackle.

Nessie thinks, 'So what comes next?' I say, 'Isn't it amazing how we got from a single cell bacterium to Paul Dirac, M. C. Escher, J. S. Bach and Scooby-Doo?'

How did it happen? 'Evolution! My little fur friends, evolution of species.'

A concordance check: is mind just a really complex algorithm?

IIT has emerged as a compelling framework within the field, offering a potential means to quantify levels of consciousness based on the integration of information within a system.

Its emphasis on brain's complexity and interconnectedness aligns with the evolutionary imperative to accurately decode physical reality for survival.

While IIT focuses on the internal mechanisms of consciousness, the concordance model in cosmology provides a complementary perspective. This model suggests an underlying order and structure to the universe, governed by physical laws and forces.

The ability of diverse brains to successfully perceive and navigate this same structured universe reinforces the notion of an independent physical reality. This shared, consistent reality shapes the evolutionary development of consciousness across the vast spectrum of neural network-based organisms.

In a nutshell

IIT suggests consciousness is a product of information integration within the brain.

The Evolutionary Lens draws inspiration from IIT of consciousness, particularly its emphasis on brain's capacity to process information, however, it takes this further.

It proposes that consciousness evolved as a tool specifically for decoding the nature of physical reality. This implies survival and reproduction are best served by an accurate perception of reality.

Brain may be susceptible to cognitive biases and influenced by cultural and societal factors, construct a version of reality that is not always congruent with the mathematical and physical descriptions of the external world.

We will shortly show that brain is a statistical modelling tool which means it can get it wrong but, as we will see later, much, much less wrong than physics and mathematics.

By saying *brain* rather than *the brain* we suggest a more abstract or conceptual approach. We imply a broader perspective, not necessarily limited to specific instances or organisms. Our exploration into the potential for consciousness extends beyond the biological realm. In future chapters, we'll delve into artificial intelligence and the simulation of consciousness.

Chapter 3:
Conservation Of Genetic Information

Evidence suggests that even seemingly simple creatures possess a spark of consciousness, reminding us that humans are not the sole proprietors of awareness.

This chapter delves into the rise of consciousness from the process of the evolution of species. We discuss the climb from single-cell organisms to conscious beings; the power of the arms race of evolution.

Evolution unfolds through a physical process, the conservation of genetic information. This principle ensures that the blueprints for life, carried within our DNA, are meticulously copied and passed on. But this isn't just about physical traits; conserved information also plays a significant role in shaping our conscious experience.

I often ask my goofy collaborators how did they ever survive natural selection?

My collaborators and I are made up of collections of cells, each collection with specialist competencies. Our cells were once unicellular organisms with all the skills needed to survive in a competitive and complex world.

The journey that we all took, those progressive steps by which we construct ourselves, our bodies and our brains are maybe the most insightful discovery of all of science.

We all start life as an unfertilised cell in an ovary, oocyte, and then slowly, step by step that oocyte turns into a collection of cells that self-construct an embryo. Eventually that embryo matures and becomes a large-scale adult; an individual with cognitive capacity and ability to reason.

All four of us have our origin in physics and chemistry of that oocyte. The *magic* (it is really not magic; we just find biology awesome!) of developmental biology is that there is a mechanism by which all these cells get together and are able to cooperate towards large-scale goals.

We are all a multi-scale competency architecture: Cells → Tissues → Organs → Bodies → Consciousness. Each of these layers has certain problem-solving capabilities in their own space.

Cells are simultaneously solving problems in physiological space, metabolic space, gene expression space, etc.

For example, during embryogenesis or regeneration, cells are solving problems in anatomical space. They navigate a path from the shape of an early embryo or a fertilised zygote, a fertilised egg cell that results from the union of a female gamete (egg, or ovum) with a male gamete (sperm), all the way to the complexity of a doggo or human body, with all of the different types of organs and structures.

Within each of us, Nessie, Iesha, Lilly and me, every cell acts as a *competent agent*, possessing its own *preferences* and *goals*, actively working to achieve them.

What evolution has given us is an interesting top-down integrated system view where every level shapes the behavioural landscape of the levels below, and the levels below do clever and interesting things that allow the levels above not to have to micromanage.

Darwinian theory of evolution

Charles Darwin's theory of evolution by natural selection[1] is one of the most influential theories in the history of science. It explains how life on Earth has changed over time. It was first proposed in his book *On the Origin of Species* published in 1859.

There are plenty of excellent resources that delve into the theory, for our purpose we can summarise the theory as such:

All living things share a common ancestor and have evolved over time through natural selection, where advantageous traits increase an organism's chances of survival and reproduction.

Variation within populations, caused by genetic mutations and environmental factors, is the foundation of evolution.

Traits are *inherited* from parents to offspring and adaptations, beneficial traits that improve survival, arise through natural selection.

Speciation, the formation of new species, occurs when populations become isolated and evolve different traits, leading to the diversity of life on Earth.

The evolutionary process is wasteful of life and energy and does not necessarily result in the most optimised solution, but it does lead to solutions that aid survival of more adaptive species.

Organisms which do not possess advantageous traits die off. An estimated 90–99% of all species that have ever existed are now extinct. Although the total number of species on Earth is uncertain, it is likely that billions of species have gone extinct since life first arose.

Natural selection has played a significant role in their extinction.

Natural selection is a manifestation of self-organisation but at a very slow time scale. Success is just existing and existence is maintaining organism's state over sufficient amount of time to propagate. This is success. For a single cell, it is from hours to days and months, in the case of our species much longer.

Darwin's Origin of Species by Means of Natural Selection maps our journey from bacteria to furry collaborators!

The Selfish Gene

Richard Dawkins, in his revolutionary book The Selfish Gene (1976), explains the theory of evolution in terms of the law of conservation of genetic information.

This is a powerful representation of the Darwin's evolutionary law in mathematical terms giving us a formal platform for building our Evolutionary Lens.

The theory of the selfish gene is a gene-centred view of evolution; in fact, it is a genetic information conservation view of the evolution of species. It describes genetic information, rather than individual organisms or groups of organisms, as the fundamental unit of selection.

Genes are the units of information that are passed down from generation to generation and are ultimately responsible for the evolution of life.

He writes, "… genes are *selfish* because they are always trying to replicate themselves to conserve genetic information. They do this by ensuring that the organisms that carry them are successful at surviving and reproducing."

Genes that are successful at replicating themselves tend to spread through the population.

The selfish gene theory is really a metaphor. It is a way of thinking about evolution that helps us understand how it works, a powerful and influential way of understanding the nature of life and the forces that shape it.

It is not a literal description of how genes work. Genes do not have intent. They are molecules that often code for proteins. They provide the blueprint for creating the essential building blocks of our bodies.

Consciousness is explained by the selfish gene theory as a powerful tool for organisms to make better decisions about how to survive and reproduce by taking into account a wider range of factors, such as the behaviour of other organisms, the environment and predict the future.

What does this genetic information look like? How does it result in such amazing species diversity while also contributing to the brutal annihilation of exponentially more species?

Conservation of genetic information

Nessie, Iesha and Lilly share about 82% of their DNA code with me. This means that 82% of the genes in my genome are also found in their genome. The remaining 18% are responsible for the many physical and behavioural differences between us.

We share a common ancestor that lived about 40 million years ago. Over time, our DNA has evolved differently, but we still share many of the same genes, the most important of which is the gene responsible for the love of cheese!

Information is different from other concepts in fundamental physics. Unlike observables such as a particle's charge or momentum, it is intangible.

The consistent patterns in the laws of physics allow information to exist in our universe.

Laws of physics govern processes such as computation, communication, as well as the existence of living organisms that inherently carry information about their own design.

Lilly looks puzzled. 'Are computation and information governed by laws of physics? Not mathematics and logic?' 'Yes!' I explain. 'Isn't it amazing how quantum computation has completely revolutionised our understanding of information and computation?'

Modern synthesis of evolution combines Darwin's theory of natural selection with the principles of genetic information. It explains how genetic variation is generated, how it is inherited and how it is acted on by natural selection.

Darwin's evolutionary processes in terms of genetics information processing

Living organisms are made of atoms and their design is coded within their DNA sequence.

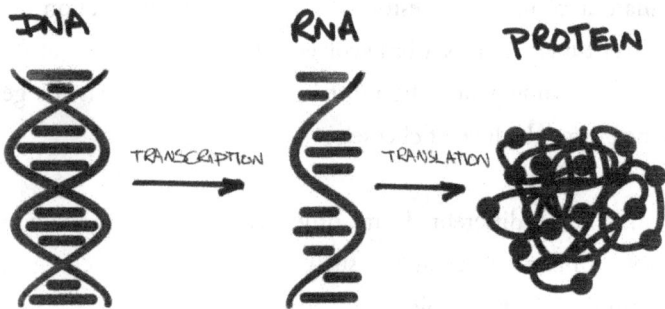

Figure 6: Genetic information flows from DNA to RNA to proteins, ultimately determining the traits and functions of an organism

Within any population of organisms there are subtle variations in their genetic composition or DNA. These variations, known as mutations, can arise spontaneously, such as errors during DNA replication, or they can be induced by environmental agents.

These genetic alterations serve as the foundation for evolutionary processes. If a mutation enhances an organism's chances of survivaland reproduction, it is more likely to be passed on to subsequent generations. This phenomenon, termed natural selection, leads to the gradual accumulation of advantageous traits within a population, while detrimental ones diminish. In this manner, organisms adapt to their surroundings over time.

Occasionally, these genetic changes can be so substantial that they give rise to entirely new species. Consider a scenario where a population of animals is geographically separated by a natural barrier, such as a mountain range or a river. Over extended periods, the isolated groups may diverge genetically to such an extent that they become reproductively incompatible, resulting in the formation of distinct species, a process known as speciation.

The narrative of evolution is inscribed within our genetic material. It is a tale of inheritance, of random mutations, and of how those mutations can culminate in remarkable biodiversity. It is the account of how, over vast geological timescales, minute genetic alterations and the selective pressures of the environment have moulded every living organism on our planet.

Let's examine this mechanism through the lens of information processing.

Figure 6 reveals the following details: Deoxyribonucleic acid (DNA) serves as a repository of genetic information, containing instructions that guide the development and function of an organism. This information can be passed on during cell division through a process called replication, where DNA creates identical copies of itself.

The genetic code within DNA is transcribed into single-stranded ribonucleic acid (RNA) molecules. This process, known as transcription, allows the genetic information to be utilised within the cell. For successful replication, a cell must be capable of reproducing itself accurately.

RNA molecules are then translated into proteins, the functional workhorses of the cell responsible for carrying out diverse biological processes. In some unique cases, such as with retroviruses like HIV, RNA can be reverse transcribed back into DNA.

DNA functions as both a data repository, storing genetic information and a code, containing instructions for its own replication.

Shannon and information as uncertainty

Claude Shannon,[2] the father of information theory, revolutionised how we think about information itself. His seminal work in the 1940s established a mathematical framework for measuring and quantifying information.

His ideas find profound application in understanding the conservation of genetic information. DNA sequences can be viewed as repositories of biological information.

These sequences carry within them instructions about an organism's structure, function and behaviour. Information in this context encompasses both the specific arrangement of the DNA code and the probability of various sequences or mutations occurring.

Conservation of genetic information involves not only copying this information accurately but also minimising errors caused by random mutations (noise). Evolutionary success favours organisms which minimise information loss during the replication.

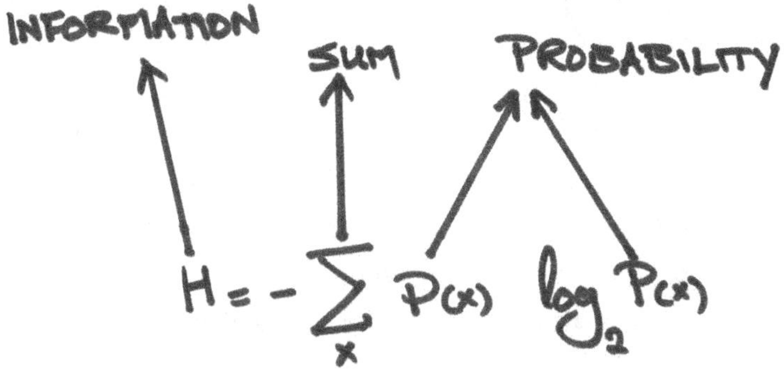

$$H = -\sum_{x} P(x) \log_2 P(x)$$

Figure 7: Shannon's information theory measures average information content (entropy)

Shannon's work helps us quantify this evolutionary drive towards *error correction* in an information-theoretic context, strengthening the link between genes and computation.

Evolution of brain

David Deutsch,[3] [4] professor of physics at Oxford University explains: "Conservation of genetic information, facilitated by the process of self-replication, explains the diverse range of adaptations observed in nature. These adaptations are the outcome of natural selection favouring knowledge disseminated throughout the population."

Natural selection, genetic mutation and environmental factors over time have favoured organisms with larger brains. Larger brains can support more complex cognitive and executive functions, such as planning, decision-making, development of language and ultimately the emergence of culture.

Our brain chemistry has also evolved over time to have higher levels of the neurotransmitter dopamine which is associated with motivation, reward and learning.

The brain's evolution over hundreds of millions of years reflects a complex interplay of natural selection, genetic mutation and environmental factors. It is a complex organ that requires much energy to develop and maintain. Natural selection favours organisms with brains that are both efficient and effective.

In the previous chapter we said that consciousness in neural network-based biology exists on a continuum rather than in distinct categories. There is no evidence to suggest we should draw an arbitrary line on the tree of animal life for the emergence of different types of consciousness.

Animals have shown they can exhibit some of the same features of consciousness as humans, solve problems, learn from their experiences and use tools, be aware of and responsive to surroundings, find food and mate to conserve their genetic code.

Koch and Tononi say that all animals experience consciousness, because they all have the ability to integrate information. The level of consciousness experienced by an animal depends on the complexity of its brain and the level of integrated information processing in brain.

In the previous chapter we saw, in contrast to us humans who have about 86 billion neurons, fruit flies have about 130,000 neurons. However, fruit flies are still capable of complex behaviours, such as learning, memory and navigation.

Information processing: brain's key to survival

Fruit flies have been often used to study the role of neurons in learning and memory, because of the small size of their brain and the simplicity of its architecture. Effects of aversive stimuli on fruit flies have been studied.

I must say, I did not discuss this with my fur friends because aversive stimuli are really unpleasant and harmful.

Studies suggests that synapses are as important for learning and memory than the actual size of brain. Fruit flies who have gone through such studies have more synapses, or connections between neurons, in their brains.

Planarians, flatworms that have a brain with only a few thousand neurons, nematodes, roundworms with a simple brain with only about 300 neurons, cnidarians, animals that include jellyfish, corals and hydras also with small brains of only a few hundred neurons, are all capable of stinging their prey, capturing food, mating, learning, memory and swimming.

Koch and Tononi have discovered that in all animals with neurons-based brain:

- all neurons have the ability to integrate information from multiple sources;

- all neurons use the same basic mechanisms to integrate information; and

- the level of integrated information processing in brain correlates with the level of consciousness experienced by an animal.

They conclude that these similarities suggest that consciousness is a fundamental property of all animals with neurons, regardless of the complexity of their brains.

The Evolutionary Lens: exploring the notion that physical reality is best inferred by brain

We are great fans of David Deutsch. His lectures on quantum computation were an invaluable resource in our exploration of quantum information theory. He is undeniably a pioneer in the field. However, when delving into the complexities of physical reality, even the most seasoned researchers get lost in mathematics.

Here he explains: "The basic idea [is] that each of us sees a different world ... and the *miracle* is that we can coordinate and cooperate as if we all saw the same world ... that is the miracle."[5]

I explain to my fur collaborators, 'Let's remember that the seemingly miraculous coordination of neural networks is shaped by their training data. This data reflects a single, unified physical reality.'

Recall Chapter 1, looking through The Evolutionary Lens we see that the process by which brain interprets sensory information, what we perceive, experience and what is shaped and modelled in our brain, may be our most comprehensive description of a single physical reality.

Limits of perception

While brain excels at processing and integrating information, it's important to remember that it's not a perfect mirror of reality. Our senses are limited.

Think of how a dog's superior sense of smell or a bat's echolocation reveals aspects of the world that humans are completely unaware of. Furthermore, brain is shaped by millions of years of evolution. It prioritises survival, which means it may filter or even slightly distort information if it helps us avoid danger or find food more efficiently.

Cognitive biases, inherent in thought processes, *confirmation bias*, the tendency to favour information that confirms pre-existing beliefs, brain's reliance on *heuristics*, mental shortcuts that simplify decision-making, introduce systematic errors in judgement.

These limitations highlight the subjectivity of our perception and emphasise that the physical reality we experience is filtered through the lens of our cognitive capacities.

Understanding physical reality: the role of brain

Brain's predictive capabilities are a survival tool honed by evolutionary pressures.

Brain does not just passively receive information from the external world; rather, it actively constructs a model of physical reality based on integrating information from sensory inputs, internal representations, past experiences, memories and expectations.

It draws upon a vast reservoir of stored information to make sense of the present; constructing a narrative that aligns with our existing mental frameworks.

The concept of the brain as a prediction machine, championed by neuroscientist Anil Seth[6] [7] [8] in his theory of predictive processing, aligns closely with The Evolutionary Lens.

This framework states that the ability to anticipate and model the environment provides a powerful survival advantage. Brain processes our perception of physical reality, enabling organisms to react swiftly to threats and opportunities, thus enhancing their chances of genetic propagation.

This is a predictive mechanism which allows brain to anticipate and interpret the environment, actively testing the physical reality we perceive.

Crucially, discrepancies between predictions and reality drive learning.

Brain updates its models, refining our understanding of the world, a process mirroring Bayesian statistics. In Bayesian statistics, prior beliefs and new evidence are used for decision-making.

Imagine you just adopted a new dog. You're excited to learn her personality. Does that tail wag mean pure joy, or is there a hint of nervousness?

We start with a prior belief; maybe most dogs are generally friendly. Each bark, growl or tail wag is new data.

This approach lets us continually update our understanding of our dog's behaviour, refining our predictions. Were they scared by the vacuum cleaner before, but less so now?

Bayesian statistics helps us quantify how much we should update our beliefs, making us better equipped to understand our new companion.

The ability to anticipate, model and correct provides a powerful evolutionary advantage, enhancing an organism's chances for survival and passing on its precious genetic information.

However, sensory limitations and internal biases mean brain's construction process is imperfect, underscoring the subjective nature of our perceived reality.

The illusion of objectivity

Philosophical debates about reality, from *idealism to empiricism*, converge on the question of whether an external reality exists at all beyond our subjective perceptions.

Iesha is looking puzzled. 'Are there people who say physical reality may not exist?' I explain, 'Yes and we do not agree with them. We say brain understands physical reality. It must do to survive.'

Organisms cannot survive and replicate their genetic code if they get physical reality wrong. One false move by the squirrel and she is some dog's meal. The physical reality she perceives must be *universal* and accurate. By 'universal' we mean the dog's reality is the same as the squirrel's.

Initially Nessie, Iesha, Lilly and I thought that this book was going to be about how reality is physics and it is described by mathematics. And that was all there was to it.

But gently, my fur friends started to chip away at my biases. The laws governing the micro and macro world are conflicting. Our understandings of the two physics contradict each other. Which one is *the* physical reality the dog and squirrel perceive?

Roger Penrose says, "There is something wrong with our physics."

Consciousness and self

Our proposition that physical reality is best captured by brain has serious implications for our understanding of consciousness, self-awareness and reality.

If brain actively constructs a model of physical reality based on sensory inputs and internal models, then consciousness can be viewed as the ongoing process of generating and updating these mental constructs.

The narrative of the self, shaped by memories, emotions and personal history, is not a fixed entity but a fluid and evolving representation. So, brain's role in representing physical reality is the formation of a coherent and continuous sense of self.

The power of Bayesian statistics

Bayesian statistics offers a powerful tool for understanding how the brain makes predictions. Imagine it like this: brain has initial beliefs about the world (the prior) and gathers sensory data (the evidence). It uses this evidence to refine its beliefs, like a detective updates their prime suspect based on new clues.

This model aligns with the principle of evolutionary fitness: organisms that most accurately predict and adapt to their environments have the best chance of survival.

It is, essentially, a mathematical formula that combines our prior belief, the prior distribution, with the evidence we have seen, the likelihood function, to update our belief about the situation. The result is called the *posterior distribution*, which represents our updated, more informed belief about what's going on.

I usually let Nessie, Iesha and Lilly lead the walk. Think of how they navigate a familiar route. As soon as we step out of the car, their brains hold a belief about how to choose the best route to take – smells, walkers, fur friends, humans generous with their treat pots; these are their prior beliefs.

Along the way, they encounter new information: new smells, notes left on the side of the path by their friends, a tempting shortcut. Based on this new evidence, their brains update its beliefs about the best route. This refined understanding of what's likely to happen and how to choose their path is similar to the Bayesian posterior distribution.

Just like their brains aim to efficiently reach slices of cooked chicken carried by their favourite human, Anthony, find fur friends, explore new areas, Bayesian statistics is used to reach the most probable conclusion based on data.

Brain's ability to predict and adapt based on experience is what allows organisms to survive and thrive in a constantly changing world. It constantly weighs prior knowledge against new evidence, aiming to achieve the most accurate and useful perception of reality.

Bayesian statistics offers us a mathematical framework that mirrors this process of how knowledge and belief evolve.

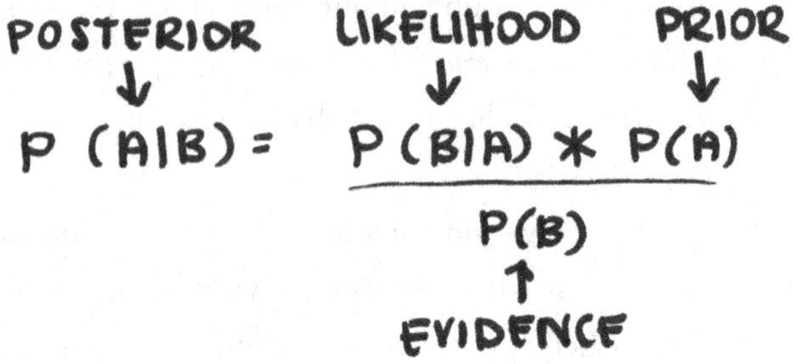

$$\underset{\text{POSTERIOR}}{P(A|B)} = \frac{\underset{\text{LIKELIHOOD}}{P(B|A)} * \underset{\text{PRIOR}}{P(A)}}{\underset{\text{EVIDENCE}}{P(B)}}$$

Figure 8: Bayes' theorem updates beliefs based on new evidence

While the pursuit of physical reality remains a scientific endeavour, the acknowledgment of the active role brain plays in understanding reality calls for a nuanced appreciation of the interplay between perception, cognition and the nature of the world we inhabit.

In the next chapter we discuss how universality of computation and Hebbian learning can lead to emergence of consciousness in neural networks; biological or synthetic.

A concordance check: why evolution of species is a brutal bug fixer

The principles of the conservation of genetic information find compelling support in the concordance model of cosmology.

This model, grounded in observations and the laws of physics, paints a picture of a universe with consistent fundamental rules and constants. These universal rules govern the behaviour of matter and energy, from subatomic particles to galaxies.

The existence of life itself, with its reliance on complex biological processes, is contingent upon the stability and consistency of the physical laws that govern the cosmos.

Genetic information, encoded in DNA molecules, relies on predictable chemical interactions and physical processes. Changes in the fundamental physical constants, such as the speed of light or the strength of the electromagnetic force, would likely have catastrophic consequences for the delicate balance of biological systems.

The concordance model suggests a universe where the laws of physics are consistent throughout space and time.

This consistency provides a stable foundation for the complex processes of replication, mutation and selection that underpin the conservation of genetic information.

Evolution, which has shaped the diverse range of life forms on Earth, is ultimately grounded in the fundamental principles that govern the physical universe.

In a nutshell

Evolution of species and the development of consciousness are the outcomes of conservation of genetic information principle.

Genetic information encompasses instructions for the development, functioning and reproduction of organisms, dictating traits, behaviours and adaptations across generations.

Looking through The Evolutionary Lens we can see that physical reality drives this diversity, variation, survival and replication.

Chapter 4:
Is The Universe Fine-Tuned For The Emergence Of Life And Mind?

Do aliens have their own furry friends?

In some quantum mechanics interpretations, the act of measurement can influence the observed state of a system, highlighting the intricate relationship between the conscious observer and the observed (Chapter 9).

In this interpretation, what happens to physical reality influenced by a conscious observer, an alien, millions of light years away from Earth? Is physical reality, in the alien's locale, different from ours?

To answer, in this chapter, we ask whether the universe exhibits conditions uniquely suited for the emergence of life and consciousness. We will explore how the fundamental laws of physics may have played a pivotal role in the development of life as we know it.

While the fundamental structure of our universe might seem surprisingly conducive to life, it doesn't guarantee the existence of mind. We suggest life's remarkable tenacity means it tends to emerge whenever and wherever conditions allow, without requiring a perfectly designed universe.

Emmy Noether's brilliant theorem links mathematical symmetries to physical laws, including conservation of information.

Her theorem suggests that the very mechanisms of evolution could be a direct consequence of the most basic laws of physics and information; implying it could happen anywhere in our universe. Are these laws directly driving evolution, implying its universality?

Is universe tuned for life?

Physics demonstrates a profound connection between gravity, subatomic particles and the basis of life itself.

The question of whether our specific universe is conducive to life and mind is an interesting one. Did we hit a jackpot, with the cosmos producing just the right conditions for us? Or, could life and consciousness emerge in a universe with a different set of rules and constants?

During our walk today, Nessie, Iesha, Lilly and I delved into the concept of fundamental constants in nature. These constants, numbering 26 in total, are essential values that govern the behaviour of our universe.

They seem to be selected specifically to allow for the existence of life as we know it. Even small alterations in their values could result in a vastly different universe, potentially one where life as we understand it would not be possible.

The precise values of fundamental constants in the universe, a phenomenon known as *fine-tuning*, have garnered attention from both religious and scientific communities.

Analysing this concept is crucial for assessing its significance in both faith-based arguments for a creator and in theoretical physics, such as the multiverse hypothesis.

Nessie thinks, 'Oh, oh, is this about Zeus?' I assure her, 'We argue against all non-scientific ideas here.'

The strong nuclear force is responsible for binding the nuclei of atoms together. If this force were too strong, all the hydrogen in the universe would have fused into helium prematurely. Which would result in the absence of stars and planets. Conversely, if the force were too weak, atoms would not have been able to form initially.

The weak nuclear force plays a crucial role in processes such as radioactive decay and nuclear fusion within stars. If this force were too strong, stars would exhaust their fuel rapidly. Conversely, if it were too weak, stars would be unable to undergo hydrogen fusion, leading to the absence of elements heavier than hydrogen.

The electromagnetic force governs the interactions between charged particles, such as electrons and protons. If this force were too strong, atoms would become unstable. Conversely, if it were too weak, atoms would be unable to form molecules, rendering life impossible.

The extraordinary precision of the Big Bang

Physicist Roger Penrose[1] [2] acknowledges the precision of fundamental constants for enabling life's existence. However, he proposes an even more extraordinary concept: that the initial state of the universe itself may possess an unparalleled level of uniqueness.

He proposed a fascinating perspective on the Big Bang. Considering the entire universe as a colossal black hole enabled him to estimate the universe's *entropy* at its origin, revealing an extraordinarily low value essential for the emergence of life.

In Chapter 11 we explore the concept of entropy, a measure of disorder. In this context, entropy signifies the initial conditions necessary for the emergence of complex structures.

According to Roger Penrose, the universe began in a state of extraordinary order, defying typical expectations of chaos. This initial entropy required a level of precision of one part in $10^{10^{123}}$; a number so immense it exceeds the estimated number of particles in the universe and can't even be written down.

This level of precision is necessary for a specific value in Einstein's theory of gravity, known as the cosmological constant, to allow for a universe like ours with the potential for life to be born.

Penrose suggests that the curvature of spacetime in the early universe had to be incredibly specific. This curvature, influenced by gravity and energy density, played a crucial role in the subsequent formation of stars, galaxies and ultimately the conditions necessary for life.

Fine-tuned or fine as is?

Throughout the observable universe, the laws of physics exhibit striking uniformity. The behaviour of light, along with the interactions of subatomic particles, appears governed by the same fundamental principles across vast distances.

This consistency raises a key question: why do these laws possess the specific values that allow for the existence of complex structures?

While our universe seems conducive to our existence, it's important to recognise that this may be a matter of chance rather than a predetermined necessity. Altering the values of fundamental constants could potentially yield a universe incapable of supporting life as we know it, or one where entirely different forms of life dominate.

What is wrong with the fine-tuning argument?

Nima Arkani-Hamed,[3] professor of particle physics at the Institute for Advanced Study, argues that "fine-tuning is a sign that we do not yet know".

He continues, "Every time we thought there was a fine-tuned parameter in physics, such as the speed of light (c), Coulomb's constant (k), the gravitational constant (G), Boltzmann's constant (k_B), etc. advancements in theoretical physics highlighted limitations in our understanding."

In our effort to understand our universe, we encounter fundamental constants: seemingly fixed values that underpin the laws of physics and shape reality. From the strength of gravity to the mass of the electron, these constants have provided a reliable framework for scientific exploration.

However, as we drill deeper, an interesting possibility emerges: what if these *constants* aren't as constant as we initially assumed? Recent advancements in theoretical physics hint at the potential for some of these values to be dynamic.

For example, the case of the universe's initial expansion rate highlights the complexities of assessing fine-tuning. Initially the rate was estimated at one part in 10^{-60}, a seemingly incredibly specific value, later calculations within the framework of general relativity yielded a probability of 1!

It turns out this parameter does not represent an exceptionally rare occurrence within the possible range of values allowed by the theory at all.

Our existing theories may be incomplete approximations of reality, requiring a deeper level of explanation for the extraordinary precision observed in fundamental constants.

Nessie, Iesha, Lilly and I are simply not convinced that there is a fine-tuning problem. The concept raises a fascinating point about the potential for life under different physical conditions in the universe.

While our specific universe exhibits conditions oddly suited for life as we know it, adjusting fundamental constants and initial conditions could indeed lead to vastly different local environments. However, the possibility of life's existence under such diverse circumstances cannot be readily dismissed.

How do we know the values we observe are unlikely to occur? Is there a way of quantifying their probability?

We think that we will never be able to work out the probability because we just do not see a constant of nature that has a different value. To work out a probability, we have to know all possible values the variable can take to determine a probability distribution empirically.

But we do not have an empirically supported distribution for the constants of nature because in our universe ... they are *constant!*

Furthermore, if we expect the universe to be fine-tuned for life, then we anticipate just enough fine-tuning and not an excess of it. However, it appears that our universe exhibits numerous instances of fine-tuning beyond what is necessary for the basic functions of life to exist.

For example, the entropy of the early universe is significantly lower than what is required to allow for life. So, what is going on here?

Conformal Cyclic Cosmology (CCC) explanation for fine-tuned parameters

CCC is a theory proposed by Roger Penrose that challenges traditional views on our universe's origin and fate. Instead of a single beginning, CCC suggests an endless cycle of expansion, contraction and renewal, interconnected by mathematical conformal transformations.

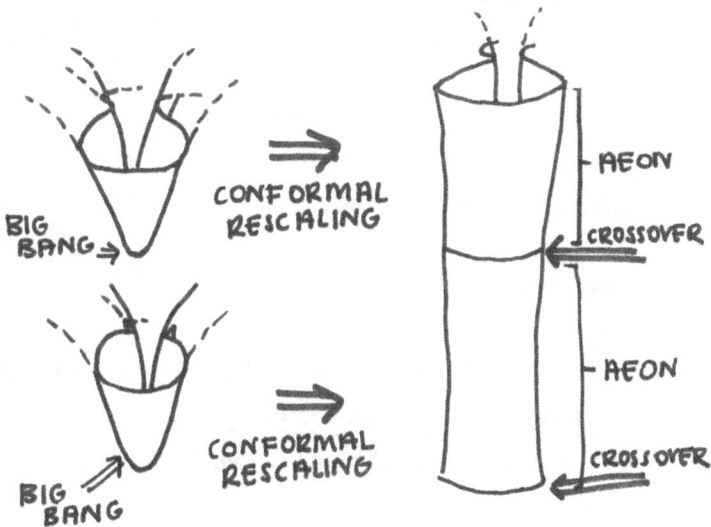

Figure 9: Cosmic Conformal Model

CCC relies on the idea that the universe goes through repeated cycles, with conformal transformations smoothing out the apparent singularities that typically mark the end of each cycle.

Unlike conventional models, CCC provides an alternative explanation for the Big Bang singularity problem, proposing that spacetime collapses after 10^{100} years leading to the restart of the cycle of universe birth.

As visualised in Figure 9, Penrose suggests that during each cycle, the universe expands and entropy increases. Eventually, matter decays into photons, resulting in a low-entropy state akin to its initial conditions. Through conformal transformations, this sets the stage for the birth of a new universe cycle.

The cyclic model of the universe proposes that physical laws and constants might change or reset across vast expanses of time.

Within this perspective, our current set of laws and constants represents just one iteration within a potentially infinite repeating cycle. This one happens to have the right constants, so we are here to discuss all the other possibilities.

Unveiling the exoplanet zoo

My collaborators think, 'If our universe is finely tuned for life, is it also jam-packed with life?' I respond, 'Oh, you are asking if there are extraterrestrial squirrels in the universe? Or could there be alien dogs in the universe?'

We participate in a popular project, Exoplanet Watch, which uses data from NASA's Kepler Space Telescope to search for transiting exoplanets. To participate, we simply registered for an account and started analysing data.

The best part? You get to name your own planet.

I must admit, I do most of the work; Nessie, Iesha and Lilly just come back from walks and fall asleep!

Only three decades ago, the notion of planets orbiting distant stars was pure speculation. The idea of a universe teeming with exoplanets remained a distant possibility in astronomical research.

Then, in 1992, everything changed. The discovery of a pulsar planet, a celestial oddity bound to a neutron star, changed the paradigm. While not hospitable for life as we know it, it proved that planets could indeed exist beyond our solar system's cosy confines.

Just three years later, another breakthrough shook the astronomical world. 51 Pegasi b, a gas giant whipping around its sun in a mere four-day sprint, became the first confirmed exoplanet orbiting a Sun-like star.

This discovery opened the floodgates: planets outside our solar system were not cosmic rarities but common in our universe.

Astronomers, armed with tools such as the Kepler Space Telescope and the Transiting Exoplanet Survey Satellite (TESS), began discovering exoplanets.

Using techniques like the dimming of starlight during transits and the subtle gravitational tug-of-war detected in radial velocity measurements, 5,528 confirmed exoplanets in over 4,000 planetary systems have been identified.

Many of these systems, harbouring multiple planets, offer miniature solar system landscapes waiting to be explored.

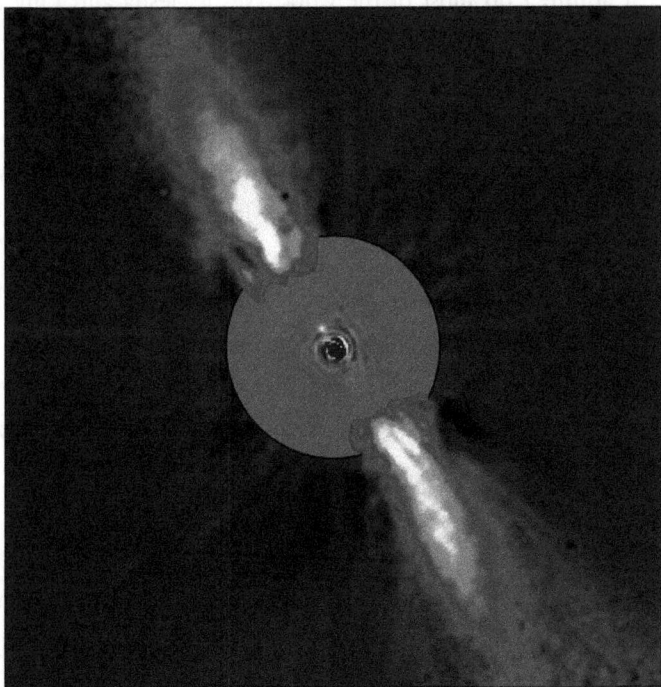

A team of French astronomers used the European Southern Observatory's Very Large Telescope to discover a giant exoplanet, Beta Pictoris b, resulting in this iconic direct image. Image credit: ESO/A.-M. Lagrange et al.

The sheer number of discovered exoplanets dwarfs the cautious speculation of just a generation ago. The universe, once the domain of galaxies, stars, planets and black holes, now includes countless alien worlds.

Understanding the diversity and intricacies of these exoplanets has become the new grand astronomical project.

The future beckons

JWST, with its capacity to peer into exoplanet atmospheres, promises even more revelations.

The hunt for Earth-like worlds, with potential for life, intensifies. Each new discovery paints a more detailed picture of our universe and its potential for life beyond our 'pale blue dot', as Carl Sagan famously put it.

We tend to think of the universe as what we see in the sky, but there's actually much more to it. Our observation tools and satellites can only show us a part of the universe that light has had time to reach us from.

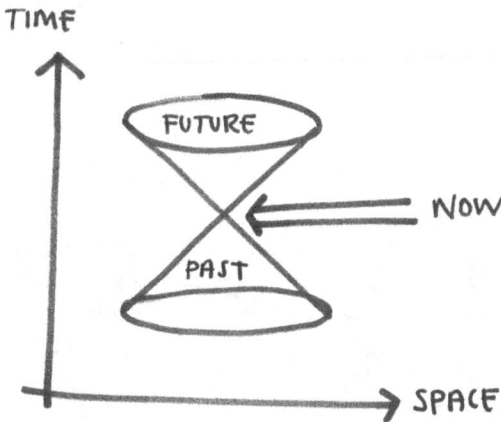

Figure 10: Our light cone

What we actually see is the path light takes in spacetime, a four-dimensional construct that combines the three dimensions of *space* with the dimension of *time* in Figure 10. This path creates a cone-like structure called a light cone.

Our Past Light Cone: The past light cone represents all the events in the universe that could have possibly influenced us. It contains all the light that has reached our eyes up to the present moment. When we look at a star, we are seeing it as it was in the past, because the light from that star has travelled an enormous distance across space and time to reach us. The farther away the star, the further back in time we are seeing it.

Our Future Light Cone: The future light cone represents all the events in the universe that we can potentially influence. It contains all the possible paths that light emitted from us right now could take as it travels into the future.

Outside the Light Cone: Events outside our light cone are causally disconnected from us. They are either too far away for light to have reached us yet, or they are so far in the future that light from us hasn't reached them. This means we cannot see them and they cannot affect us.

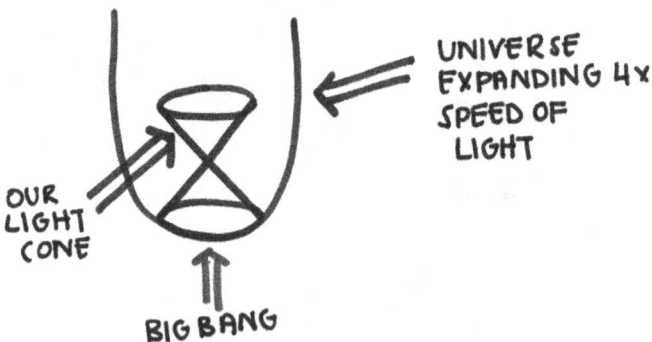

Figure 11: Visible universe

Therefore, our observations and discussions in this chapter will be confined to our observable universe, the region encompassed by our past light cone.

I explain to my collaborators, 'Just a quick diversion whilst we are on this topic, the Cosmic Microwave Background (CMB) radiation, a faint afterglow of the Big Bang, is entirely contained within our past light cone. The CMB radiation has been travelling for approximately 13.8 billion years to reach us. This is because it originated about 380,000 years after the Big Bang. How cool is this? We can see the remnant of the Big Bang on an analogue TV or hear it on an analogue radio!'

The Goldilocks zone, also known as the habitable zone, is the range of distances from a star where liquid water could exist on the surface of an orbiting planet.

Liquid water is considered crucial for the possibility of life based on our current understanding, making the Goldilocks zone an important concept in the search for potentially habitable exoplanets.

The exact boundaries of the Goldilocks zone vary depending on the type of star. For example, the Goldilocks zone for a red dwarf is much closer to the star than the Goldilocks zone for a Sun-like star.

This is because red dwarfs are cooler than Sun-like stars, so their planets need to be closer to the star in order for liquid water to exist on their surfaces.

Some exoplanets discovered are Earth-sized, which makes them interesting to explore. Examples include Kepler-186f, Kepler-

452b, Proxima b, TRAPPIST-1e and Teegarden b. All of these are considered potential candidates for habitability.

Drake's equation

Let's cut to the chase, are we alone in the universe, or is life a common occurrence in our universe?

Drake's equation, developed by American astrophysicist Dr Frank Drake in 1961, is a probabilistic formula used to estimate the number of extraterrestrial civilisations in our galaxy.

$$N = R\star \; x \; fp \; x \; ne \; x \; fl \; x \; fi \; x \; fc \; x \; L$$

Let's delve into the meaning of each variable:

1. *N:* The number of civilisations with which humans could communicate in our galaxy
2. *R*:* Average rate of star formation in universe
3. *fp:* Fraction of stars that have planetary systems
4. *ne:* Number of planets that could potentially support life per star with planets in the habitable zone
5. *fl:* Fraction of planets where life actually emerges
6. *fi:* The fraction of those planets with intelligent life.
7. *fc:* Fraction of planets with intelligent life that develop technology capable of interstellar communication.
8. *L:* The length of time those civilisations release signals into space

This formula has been instrumental in shaping the Search for Extraterrestrial Intelligence (SETI) programmes and guiding the scientific community's efforts to explore for life.

One of the primary challenges lies in determining the values for the variables, many of which remain unknown or are difficult to quantify precisely. Nevertheless, it remains a very useful tool for our purpose.

Thousands of exoplanet discoveries are a treasure trove of data, giving us real numbers to plug into some of those Drake equation variables, such as *fp* and *ne*. Advances in astrobiology have also offered insights into the conditions necessary for life to emerge *(fl)*.

Is the universe old enough for emergence of conscious life anyway?

The Sun and the Earth appeared 4.5 billion years ago, and it took another 5 billion years for environmental conditions to be right for the proliferation of life. Then, evolutionary processes did their work to lead to the emergence of brains and consciousness.

Current estimates are that conscious life takes at least 10^{10} years, *Darwin Time*, to emerge. Given the universe is only 1.4×10^{10} years old, that is 14 billion years, it appears we have arrived early on the scene.

N may be a small, but non-zero, number.

Nessie, Iesha and Lilly think that if N is so tiny, then the three of them must be close to the pinnacle of universal cleverness. Looking at my goofy friends, I must disagree.

Atmospheric, surface and temporal life signatures on exoplanets, detectable signs of life beyond Earth, can help us fine-tune Drake's equation.

JWST, launched in 2021, is capable of detecting oxygen, methane, nitrous, chlorophyll, infrared light and seasonal changes, which are among the most promising life signatures.

With each new exoplanet discovered and every breakthrough in astrobiology, we are improving our ability to calculate N. Multiplying N by the number of galaxies that can harbour life in the visible universe gives us a clue about the proliferation of life in the universe.

While none of these signatures alone provide definitive proof of life on an exoplanet, the detection of multiple life signatures on the same exoplanet would constitute strong evidence of the presence of life.

Now that we've discussed the potential existence of life in the visible universe, let's shift gears and talk about Emmy Noether.

But before Emmy Noether, for the sake of completion, here is the doggo formula to estimate the number of alien civilisations in our galaxy who carry treats for their alien dogs. As you can see, we had a productive walk this morning:

$$N = R* \times fp \times ne \times fl \times fd \times fc \times fi \times L$$

Where:

1. *N:* The number of treat-carrying civilisations in our galaxy!
2. *R*:* Average rate of star formation in the universe
3. *fp:* Fraction of stars that have planetary systems
4. *ne:* Number of planets that could potentially support life per star with planets in the habitable zone

5. *fl:* Fraction of planets where life actually emerges
6. *fd:* Fraction of planets where dogs emerge
7. *fc:* Fraction of planets with intelligent life that develop technology to create cheese
8. *fi:* The fraction of planets with intelligent enough life to know dogs like treats
9. *L:* The length of time those civilisations create cheese

The law of conservation of information

Emmy Noether, a brilliant mathematician, left a profound legacy in theoretical physics. Her theorem, formulated in 1918, unveiled a deep and intricate relationship between symmetry and conservation laws.

Noether's Theorem, now considered a cornerstone of modern physics, connected the abstract realm of mathematics with the concrete laws governing the conservation of energy, momentum, angular momentum and information.

It is impossible to overstate Noether's genius.

In physics, *symmetry* refers to the *invariance* of a physical system under specific *transformations*.

Invariance means that there is no change to the properties of the system if you move the system in space or time.

Transformation is the 'change' applied to the system. It could be moving the system in space, making it bigger or smaller or even observing the system from a different perspective.

Symmetries in the laws of physics suggests they are valid throughout our universe.

Noether's Theorem states that every symmetry in a physical system corresponds to a conservation law. This genius insight serves as a bridge between abstract mathematical concepts and the observable, tangible phenomena in the universe.

Energy cannot be created or destroyed, only converted between forms. This principal stems from time symmetry – physical laws work the same regardless of when an experiment occurs.

Noether's Theorem links this symmetry to *energy conservation*, a cornerstone of our understanding of the universe.

Likewise, momentum is conserved because of spatial symmetry. Physics doesn't change based on location. Noether's Theorem connects this symmetry to the *conservation of momentum*.

Angular momentum, describing rotational motion, is also conserved. This arises from rotational symmetry – physical laws stay the same under rotations, therefore angular momentum is conserved. This affects objects from spinning tops to the vast structure of spacetime.

From the symmetries of subatomic particles to the orbits of celestial bodies, the principles of conservation find their origins in the genius framework of Noether's Theorem.

In 1993, Jacob Bekenstein and Leonard Susskind[4] showed how Noether's Theorem can be extended to describe information as a conserved quantity in the universe.

Conservation of information is based on the principle that the total amount of information in the universe remains constant over time, treating information as a physical quantity akin to energy and momentum.

While the law of conservation of information does not dictate the mechanism for conservation, it simply states that information remains constant.

In the previous chapter, we presented David Deutsch's argument that the universality of computation is a physical phenomenon describing what is computable in our physical universe.

Information is a fundamental aspect of physics, underlying chemistry and biology, and consequently, all forms of life, regardless of their origin.

The universality of genetic information

This implies even if alien life has a different genetic code than life on Earth, the law of conservation of information would still apply. The genetic information of alien life would still be encoded in some form of DNA or its equivalent, and it would still be conserved from generation to generation.

> "The human brain's capacity for explanatory computational universality is encoded in a remarkably small portion of DNA – merely a minuscule fraction.
>
> While 98.8% of our DNA is shared with chimpanzees, the remaining 1.2%, specific to

humans, largely consists of junk DNA or distinctions between the two species that aren't epistemological, meaning they aren't related to the nature, origin and boundaries of human knowledge.

This distinct portion could potentially be encapsulated in just a few thousand lines of code governing human general intelligence.

We haven't discovered the epistemological DNA code yet, but we're excited about ongoing research in the Integrated Information Theory of consciousness. It offers a promising route for guiding the philosophical theory of consciousness.

Once we have high-level, precise terms for abstract concepts like creativity, consciousness, and qualia, we can then develop a program that embodies those features."

David Deutsch[5] Constructor Theory Lecture

From the simplest bacteria to complex multicellular organisms like ourselves, the elegant structure of DNA describes life.

Across the vast life form on Earth, variations in DNA sequences give rise to the incredible diversity we observe. Yet, the underlying language of nucleotides, adenine, guanine, cytosine and thymine remains astonishingly consistent.

The universality of DNA reveals a fundamental principle of evolution: genetic information acts as a universal building block within all lifeforms.

Conservation of information theorem explains similar, not necessarily identical, language exists within the whole universe.

This is getting confusing for my friends. 'Universal?' I explain, 'By universal we mean rules that apply to our universe, not just our planet.'

A concordance check: did the universe hit the jackpot for us?

Must universe spawn consciousness? Does consciousness exist anywhere else in the universe?

The concordance model provides a framework for understanding the delicate balance of forces and constants in the universe that seem necessary for life's existence.

From a purely empirical perspective the emergence of consciousness, as we experience it within our universe, appears to be an anomaly. The fundamental laws of physics operate seemingly devoid of any intrinsic need for subjective experience. Particles interact, fields oscillate and the cosmos expands, yet from this complex interplay arises consciousness.

The remarkably low entropy of the early universe seems incredibly fine-tuned for the emergence of complex structures and, ultimately, the type of life we see today. The concordance model helps us grasp the significance of this condition, and how a different initial state might have profound implications.

A perspective finding support in the field of artificial intelligence suggests that consciousness is an inevitable consequence of the integration of complex systems interacting with the physical world. As systems attain a greater level of information processing and organisational capacity, they may inherently give rise to subjective experience.

In a nutshell

We oppose popular interpretations of quantum mechanics and suggest physical reality remains indifferent to the emergence of life and consciousness (Chapter 9).

The existence of physical reality is not contingent upon our presence to perceive it.

Chapter 5:
Is Consciousness Computable?

There is no ghost in the machine

Taking our discussion from the previous chapter a step further, we now propose a thought experiment: What if the act of measurement, which famously creates the observed state of a system in quantum mechanics (we explore this in Chapter 9), is not carried out by a biological entity, be it human or alien, but by a conscious machine? Would this challenge our current understanding of physical reality?

This chapter argues that consciousness is a computable phenomenon because it is not a purely logical or mathematical construct.

Gödel's *incompleteness* and Turing's *undecidability* theorems expose boundaries in formal systems, raising questions about whether the mind can be fully understood as a computational process.

These theorems suggest a fundamental unpredictability within computational systems. If aspects of consciousness, like self-awareness or subjective experience, rely on incomplete or unpredictable processes, then perfect computational modelling of consciousness might be impossible.

While Gödel's and Turing's theorems highlight limitations of formal systems, we argue that they do not directly translate to limitations on the emergence of artificial consciousness.

Here's the thing that changed everything: computation is not maths or formal logic; it is physics

A Financial Times[1] article published on 11 January 2019 is titled: 'Can man ever build a mind?'

My collaborators, who differ from man in both gender and species, exhibit unmistakable signs of disappointment.

Let's rephrase the question for my colleagues: Why is it important to know if consciousness is computable?

If consciousness is ultimately computational, i.e. it is a product of computational processes, then our subjective experiences are not fundamentally different from the physical world around us.

This could challenge our traditional notions of dualism, where mind is seen as separate from the body.

In the 1970s, Marvin Minsky, the father of AI, declared that "the human brain is just a computer made of meat". That is, brain uses a formal system of algorithms to solve problems. Minsky's declaration led science and philosophy communities to conclude that consciousness cannot be simulated on a computer.

He referred to Kurt Gödel's incompleteness theorems and Alan Turing's halting problem as reasons why if intelligence or consciousness is algorithms, or software being run on the fleshy computer, then it cannot be simulated on computer systems.

John Searle says that "consciousness is not computable". He

believes consciousness cannot be simulated or replicated through computational processes. He argues that "consciousness is governed by the laws of classical complex formalism".

We argue that such propositions are without merit and are defeatist. *Incomputability* arguments have delayed scientific research in the field for decades, leading to a disastrous misunderstanding of information theory, incompleteness theorems and the halting problem, which have been as detrimental as dualism for the study of mind.

It is a strange life living in a post-Gödel world!

I asked my fur friends, 'How can someone with such a brilliant mind as Kurt Gödel be a Nazi?' The Treat Brigade sigh and tilt their heads.

They think, 'I should take comfort in the thought that Alan Turing, another genius and the father of universal computing, beat Gödel and the rest of the Nazis.'

Before Gödel and Turing, mathematics was thought to be the pinnacle of truth; consistent, complete and deterministic. Mathematical reasoning was considered true and reliable.

Yes, there were some paradoxes in mathematics, such as the nature of infinity or set theory's self-referencing, but these were viewed as anomalies that could be resolved through tighter rules in mathematics.

Bothered by these paradoxes, Bertrand Russell and Alfred North

Whitehead produced an epic piece of work, *Principia Mathematica*. Published in three volumes between 1910 and 1913, it aimed to lay bare the logical foundations of mathematics by deriving all mathematical truths from a few basic axioms. They were both proponents of formalism, the view that mathematics can be reduced to a formal system of axioms and inference rules.

After studying their work, Kurt Gödel published his incompleteness theorems in 1931, 20 years after the publication of *Principia Mathematica.*

Alan Turing published his proof of the halting problem in 1936.

What is a formal system?

A formal system is an abstract mathematical structure used for inferring theorems from axioms according to a set of rules known as the logical calculus of the formal system.

Such systems, also known as *axiomatic systems*, are powerful tools used for reasoning about mathematics and computer science. These systems are essential for proving theorems, designing programming languages and verifying the correctness of software.

Think of a formal system as a game with strict rules: There are starting conditions, i.e. *axioms*, and moves you can make, and a way to determine a winner. Gödel and Turing showed that such complex games have limitations.

Gödel: I am not provable!

Kurt Gödel's incompleteness theorems are two theorems in mathematical logic that demonstrate the inherent limitations of any complex formal axiomatic system.

In essence, Gödel's theorems prove that no such system can be both consistent and complete.

At the young age of 24, Gödel conceived the notion that numbers could represent symbols, allowing him to create a sentence that ostensibly pertained to numbers but simultaneously operated on a second level, addressing symbols.

Then, he mapped the entire structure of formulas in *Principia Mathematica* onto numbers. Within which, Gödel crafted the sentence: "There does not exist a derivation of a certain formula in *Principia Mathematica.*"

What he devised was equivalent to: 'I am not provable!'

He constructed a self-referential statement within the fortress that Russell and Whitehead had constructed to exclude self-reference. It circumvented all of Russell's efforts and transformed self-reference into an unavoidable phenomenon.

Gödel's brilliance in proving this theorem lies in his invention of a method to translate any complex formal system into number theory and demonstrate that number theory is inherently self-referential.

Alan Turing added to the mathematicians' misery by declaring that there is no way to guarantee that a complex enough axiomatic and algorithmic formal system will not get stuck in an infinite loop.

In other words, there is no way of guaranteeing such a system is deterministic!

Mathematics had been discovered to be inconsistent, incomplete and non-deterministic.

Gödel's first incompleteness theorem shows that there are limits to what mathematics can achieve, and that there are some truths about the natural numbers that are beyond the reach of proof. The second incompleteness theorem shows that there are limits to what logic can achieve, revealing that certain truths about the world lie beyond the grasp of reason.

Mathematics is not flawed, but limited in its scope.

Gödel and Turing do not say that mathematics is incorrect. Rather, they demonstrate that a proof system may be unable to prove a statement, although we might recognise its truthfulness if we adhere to the formal system.

Numerous misconceptions surround the Gödel theorems.

One common misunderstanding is the assertion that they prove consciousness to be non-computable.

Douglas Hofstadter, the author of *Gödel, Escher, Bach: An Eternal Golden Braid,* suggests that brain, despite being composed of inanimate molecules engaged in mere chemistry, remarkably yields not only the capacity for perception but also the ability to construct a self-model and experience consciousness. His perspective involves viewing brain not solely as a physical object, but also as a profound entity capable of generating thought, feeling and awareness.

Four cocking heads at this point are testament to the fact that this does not make sense to any of us.

Let's quickly talk about this before our next doggo meeting point.

All that Gödel theorems say is that if complex systems' axioms can be listed algorithmically, then it must be incomplete.

It says nothing about the incompleteness of systems whose axioms cannot be algorithmically listed or are simple enough to avoid self-referencing.

It simply does not logically follow that because certain aspects of arithmetic cannot be proven within recursively enumerated axiomatic systems, human brains cannot be simulated by computers and are inherently *magical.*

If you had a problem with the last paragraph, do not despair. That is the whole point. There is no connection between Gödel's theorems and the simulation of human brain. All we need to know is: it just does not follow!

Consciousness is not a formal system: deduction vs induction

In logic and reasoning, two fundamental methods stand out:

deduction and *induction*. These approaches serve as the cornerstone of scientific inquiry, philosophical discourse and everyday decision-making.

Deduction and *Induction* represent distinct modes of reasoning, each with its unique strengths and limitations.

Deduction is a logical process in which specific conclusions are drawn from general principles or premises. Deductive reasoning is the basis of mathematics and formal logic, where conclusions are certain if the premises are true and the reasoning is valid. Axiomatic formal systems use this process for proving theorems.

I think the fur girls need an example. I explain, 'Premise 1: all major roads are backed up during rush hour. Premise 2: it is rush hour right now. Deduction: the major roads are backed up.'

Oh, doggies don't like this example, it means they have to wait longer for their walk and cheese. What a sad morning!

The limitations of deduction-based systems, such as those subject to Gödel and Turing's theorems, underscore the importance of induction in understanding both consciousness and the physical world.

Induction, on the other hand, is a process of reasoning where specific observations or instances lead to general conclusions. In inductive reasoning, individuals draw conclusions based on patterns observed in specific cases.

Birds may start their lives sampling various insects, guided mostly by instinct. But through trial and error, they quickly learn to associate certain visual cues, a colour, a pattern, with a particularly tasty or unpleasant experience.

Over time, they refine their hunting techniques, focusing on those insects that provide the best nutritional reward. This is inductive learning in action.

Similarly, a squirrel that repeatedly finds its food stash raided by other squirrels might start devising new hiding strategies. Perhaps burying nuts in multiple locations, creating decoy stashes, or choosing less predictable hiding spots altogether.

This adaptation demonstrates inductive thinking, as the squirrel draws conclusions from past experiences to modify its behaviour for better survival outcomes.

Induction is valuable for generating hypotheses and making predictions, it does not provide certainty. The conclusions drawn through induction are probabilistic and subject to revision when new evidence emerges.

This is the method brain and universal computers use to solve problems that are currently unprovable or undecidable.

Induction is a powerful problem-solving strategy mirroring evolutionary processes themselves. Just as organisms learn to adapt to their environment through trial and error, inductive computational models constantly self-correct through exposure to new data.

This is seen in modern machine learning algorithms, which can identify patterns far beyond human capabilities without being explicitly programmed.

If consciousness arises from brain's computational powers, then inductive learning is key, highlighting the potential of such models to illuminate the processes behind our own subjective experience.

The claim that consciousness is non-computable often hinges on the idea of understanding or insight that cannot be reached by mere calculation. However, inductive processes like those seen in machine learning demonstrate the power of learning to extrapolate general principles from specific examples.

This ability to grasp patterns and concepts that go beyond initial programming directly challenges the notion that consciousness demands a *special* type of computation inaccessible to physical systems.

Is mind possibly quantum mechanical?

The self-referencing and self-exploration capabilities of brain are frequently cited by philosophers as reasons why consciousness may require systems different from formal axiomatic or algorithmic ones. After all, self-referential systems are often deemed incomputable and inherently inconsistent, aren't they?

Penrose[2] explains Gödel's theorem by highlighting that within a formal system, one can formulate statements that cannot be proven, yet intuitively appear true. He argues that this is understanding and not calculations.

It's not just about obeying the rules, but about understanding the reasons behind them. This deeper understanding allows us to see beyond the limitations of the rules themselves and grasp the bigger picture.

To Penrose, consciousness transcends mere neurological activity, suggesting it may involve quantum mechanical phenomena rather than being solely dependent on electrical impulses and neurotransmitter release.

This perspective challenges the notion that consciousness can be fully described and understood through computation or physical processes alone.

Quantum mechanics, however, has its own version of non-computability. According to quantum mechanics, particles can exist in many places simultaneously, as a wave of temporal development of properties of the particle, known as the *wave function*.

The properties of a particle only become known through interactions with other particles. Until such interactions occur, they remain unknown and non-computable.

Penrose and Stuart Hameroff,[3] [4] [5] professor emeritus at the University of Arizona, jointly developed a theory of consciousness which proposes that symmetrical hollow tubes, microtubules, within neurons can act as quantum computers, and that the collapse of quantum states in these tubes gives rise to conscious experiences.

Unlike others, Penrose's argument for the incomputability of consciousness delves far deeper than incompleteness theorems and halting problem.

He fundamentally questions our understanding of quantum mechanics.

As Lee Smolin put it, "No one quite knows what to make of Penrose's theory, but conventional wisdom goes something like this: Penrose is so brilliant, one of the very few people in my life whom I, without reservation, call a genius. His hypothesis must be taken seriously."

Penrose proposes that the mystery of consciousness is tied up with the mystery of quantum mechanics. Penrose-Hameroff's theory proposes a very specific role for microtubules, signal processing within the brain.

In the next chapter, viewed through The Evolutionary Lens, we will demonstrate that consciousness is independent of the mechanism of signal processing.

We propose that biological systems can evolve to harness quantum phenomena, enhancing their cognitive processes and providing a survival advantage.

This morning's wet walk was productive. Nessie hates getting her paws wet, giving us a chance to have a good chat. I tell her, 'What we have here is a *neurophysical* model which explain microtubule function in signal processing. It simply does not explain the emergence of consciousness.'

The explanatory universality of brain

Computing is a branch of physics and not mathematics. Computer science is not mathematics or formal logic.

Quantum information theory demonstrates that universal quantum computers, of which classical universal computers are just a special case, adhere to the principles of quantum mechanics.

Surprisingly, incompleteness and halting theorems are found to have no direct connection to physics.

David Deutsch, often referred to as the father of quantum information theory, is renowned for developing the first universal quantum computer program. He contends that brain, being a physical system, possesses the capability of universal computation.

This assertion is based on the demonstration that brain can be simulated by a Turing machine, which serves as a theoretical model of a universal computer. The principle of universality of computation dictates that any computation achievable by one computer can also be executed by any other computer, provided adequate time and memory resources.

Consequently, any physical system capable of universal computation holds the potential for implementing consciousness.

In his book *The Beginning of Infinity* (1999), Deutsch writes, "The universality of computation is a fundamental property of the universe and physics." He argues that this concept is pivotal for comprehending both consciousness and the potential for artificial general intelligence.

Doggies abruptly come to a halt, and stare at me. 'Have we not spent the whole of this chapter talking about incomputability? Is Deutsch saying that a universal computer can be conscious?'

I answer: 'YES! Emphatically YES!'

Incompleteness theorems and halting theorem are confined to axiomatic systems and algorithms that remain fixed and unchanging.

Deutsch, on the other hand, illustrates that universal computers have the ability to learn and solve problems that are considered not computable, to self-correct.

Deutsch argues: "Gödel and Turing showed that certain problems cannot be solved by any fixed and unchanging algorithm. However, universal computers are not static; they can learn and adapt in ways that Gödel and Turing's theorems do not apply. Consequently, universal computers will eventually have the capacity to solve problems that are currently unprovable or undecidable."

In *Computing Machinery* and *Intelligence* (1950), Alan Turing argued that a universal computer, with a relatively small program, could potentially demonstrate consciousness.

Brain, despite its complexity, is described as a highly regular machine composed of interconnected neurons, each performing simple operations. Brain's computational capability is attributed to the intricate network of these neurons.

Understanding the interconnections between neurons is the key to comprehending brain's computational capacity.

It's important to distinguish between computation as a mathematical construct and computation as a physical phenomenon within our universe. Axiomatic systems and formal logic, where Gödel's and Turing's theorems apply, deal with abstract rules and symbol manipulation.

Physical computing, on the other hand, is governed by the laws of physics. Brain, a physical system, harnesses processes like chemical interactions and electrical signalling to perform computations.

Our mathematical models of computation have limits. Brain, as a biological computational system, is not bound by the theoretical limitations of formal logic. Think of it like a plane's autopilot; perfect flight equations don't guarantee a perfect autopilot in the real world.

Nessie is not convinced! 'How is this even possible?' I explain, 'There are two reasons incompleteness theorems and halting problem do not apply to simulation of mind; *deduction versus induction* and the fact that computation is physics and not mathematics, or formal logic.'

Computation is physics

I tell my collaborators, 'Computability was the first seminar I participated in for my Artificial Intelligence PhD research, and the complex concepts introduced there sparked a confusion that followed me throughout my research and beyond.'

If the very foundation of computation is flawed, what does that mean for artificial intelligence?

However, claims that consciousness is not computable often ignore the broad capabilities of computation.

Computation, at its core, is a physical process. It involves the manipulation of physical systems to represent and process information. Whether it's the interactions of subatomic particles in a quantum computer, or even the firing of neurons in a biological brain, computation is ultimately grounded in the laws of physics. This deep connection between computation and physics suggests that the universe itself might be viewed as a vast computational system, constantly processing information and evolving according to a set of fundamental rules. Exploring this link between computation and physics opens up new avenues for understanding the nature of consciousness. In theory, a universal computer could simulate any physical system, including the processes that give rise to consciousness.

By leveraging these models, we can gain valuable insights into the nature of consciousness and its relationship with the world around us.

It is unanimous, with a show of three paws and a hand, Nessie, Iesha, Lilly and I say this means dualism is false and we can stop wasting time talking about it.

A concordance check: myth debunked

The question of whether consciousness is computable has serious implications for our study of mind and its relation to the physical world.

While some argue that consciousness transcends computation, relying on non-computable aspects of mathematics and logic, we propose a different perspective.

We emphasise that consciousness is computable, and understanding its computational nature would give clarity on our appreciating nature of mind and physical reality.

Our viewpoint implies that computational models can provide insights into both subjective experience and the physical world.

Universality of computation: David Deutsch's work on universal quantum computers demonstrates that any computation can be performed by a sufficiently powerful computer operating under the laws of physics. This universality suggests that consciousness, being a physical phenomenon, may also be computable.

Computation is physics, not mathematics: Gödel's and Turing's theorems apply to formal axiomatic and algorithmic systems, but the laws of physics themselves are computable by universal quantum computers. Brain, as a physical system, is not bound by the same limitations of abstract logic.

The concordance model supports this view, emphasising a universe governed by consistent laws and forces. The ability of physics to encapsulate the behaviour of matter and energy suggests that even complex systems like brain follow rules that a universal quantum computer could potentially simulate.

Deduction vs induction: While deductive formal systems are limited, brain, like modern machine learning systems, utilises inductive processes to learn and adapt. These inductive processes provide adaptability that challenges the notion that consciousness demands a 'special' type of computation.

Implications of consciousness being computable: If consciousness is ultimately computable, this suggests:

> *Reduced role for fundamental divide:* There is no fundamental divide between brain (physics) and mind (subjective experience).

> *Potential for AI:* Advances in artificial intelligence and computational models will lead to a deeper understanding of consciousness and potentially artificial general intelligence.

By understanding these nuances, we can move beyond debates centred on Gödel's and Turing's work and focus on developing powerful computational models to gain insights into the nature of consciousness.

In a nutshell

In the previous chapter we argued that conditions for life, and even possibly conscious life, exist in the universe. We concluded that consciousness is not an inevitable outcome of the universe or a cosmic purpose. We described the set of conditions that had to occur for creatures capable of pondering their existence to emerge.

In this chapter we asked, what if consciousness is computable? Well, rather than being some magical force, it suggests that our thoughts and feelings might arise from processes within brain that are fundamentally understandable and potentially replicable.

Chapters 4 and 5 uncover a conflict between our current understanding of quantum mechanics and the physical reality it aims to describe. Later we will delve into significant obstacles that challenge the prevailing interpretation of quantum mechanics.

Chapter 6:
Can Consciousness Emerge In Computers?

We copied brains ... accidentally created minds?

In the 1980s, I nearly abandoned my research studies in AI because John Searle said that consciousness is intrinsically tied to biological system. I questioned how my work in neural networks architectures could possibly yield positive results if Searle's claims were true.

Searle doesn't claim brains are magic, but he insists the biological, messy, chemical processes inside them are essential. He'd say even if a computer brain seems conscious, it isn't really experiencing, like we do. But what if our messy inner workings are just nature's way of arriving at complex computation?

Artificial intelligence (AI) research stagnated for decades due to the influence of *biological naturalism* or *biological functionalism*, with advocates like John Searle at the helm.

He still, to this day, underestimates the potential for emergent properties to arise from non-biological systems.

Recently commercially released cutting-edge language models seem eerily human. They write poems, translate languages and hold conversations which feel real to us.

Could these abilities mean we are witnessing the dawn of artificial consciousness?

In the previous chapter we established that the key to creating artificial consciousness is in replicating brain's specific patterns of computation and its ability to learn and adapt in an unpredictable world.

In this chapter, we'll argue that a specific combination of technologies has the potential to produce something akin to the consciousness we experience.

This combination includes Artificial Neural Networks (ANN), complex structures inspired by the human brain, and Hebbian learning, a process through which neurons adapt and strengthen their connections over time.

So, what's with the surprise with their competence?

At this morning's walk we were talking about why the latest generation of language models impress us with their capabilities and human-like communication.

I explain to my fur colleagues, 'These models are actually quite accurate replicas of our brains, so what's with the surprise with their competence?'

What stands out in the latest generative AI platforms is their ability to learn and adapt. Instead of relying on rigid algorithmic programming they continuously evolve by absorbing and applying information from the data provided to them.

Their capabilities arise from the complex interplay of their interconnected nodes, ANN, rather than being rigidly programmed, allowing for a level of creativity and flexibility which is essential for generating human-quality creative content.

They excel at learning from massive datasets of text and code, enabling them to acquire a deep repository of the world and the relationships between words, concepts and ideas.

Their competences range from executing instructions thoughtfully to generating creative content, translating languages and providing informative responses to even the most open-ended, challenging or unusual queries.

ANN, due to their architecture, can continuously improve their performance by analysing new data and adjusting their internal connections. Just like us, this allows them to generalise their knowledge to new situations and tasks, making them far more versatile than traditional AI systems.

The building blocks of thought: ANN

As we explored in the previous chapter, Gödel and Turing's incompleteness theorems highlight the limitations of formal systems. However, ANN function differently. Unbound by the constraints of formal proofs, they learn and adapt through exposure to massive datasets.

This adaptability mirrors how brain make sense of the world and is crucial to the argument for emergent consciousness.

Artificial Neural Networks (ANN)

ANN represent a groundbreaking notion in the field of artificial intelligence (AI), drawing inspiration from the human brain to process information and make decisions.

ANN were inspired by a 1940s brain model when Warren McCulloch and Walter Pitts[1] introduced the concept of a simplified mathematical model of the human brain's neuron. Their work laid the foundation for the development of artificial neurons, the building blocks of ANN.

The real breakthrough occurred in 1957 when Frank Rosenblatt[2] introduced the *perceptron*, a single-layer neural network capable of learning through a supervised training process.

Despite the initial excitement surrounding perceptrons, researchers soon encountered limitations in ANN ability to solve complex problems. In the 1960s and 1970s, this led to the development of multilayer perceptrons (MLPs), also known as feedforward neural networks.

The introduction of backpropagation, a supervised learning algorithm, further enhanced the training of MLPs, enabling them to learn complex patterns and relationships.

The late 1970s and 1980s witnessed growing enthusiasm for AI and neural networks, but this optimism was short-lived. The field encountered challenges, leading to a period known as the *AI winter.*

Limited computational power, insufficient data and the lack of advanced training algorithms hindered progress in neural network research. During this time, alternative AI approaches gained prominence, relegating neural networks to the background.

The late 1980s and early 1990s marked a resurgence of interest in neural networks, driven by the connectionist paradigm. Researchers like Geoffrey Hinton,[3] Yann LeCun[4] and others introduced techniques such as convolutional neural networks (CNNs) and recurrent neural networks (RNNs).

These architectures addressed the limitations of earlier models and paved the way for breakthroughs in computer vision, natural language processing and other domains.

The 21st century witnessed a transformative era for ANN with the advent of deep learning. Enabled by powerful Graphical Processing Units (GPU) and a wealth of available data, neural networks with numerous layers demonstrated unprecedented capabilities in image recognition, speech processing and language understanding.

Neurons, the basic units of our brain, communicate through electrical and chemical signals, transmitting information through synapses, where neurotransmitters bridge the gaps between neurons.

A perceptron models a simplified neuron: It receives inputs (like dendrites), processes them based on weights and a bias, and generates an output (like an axon).

Picture millions of tiny switches (neurons) taking in info, then based on how they're wired, firing on or off. That's essentially an ANN.

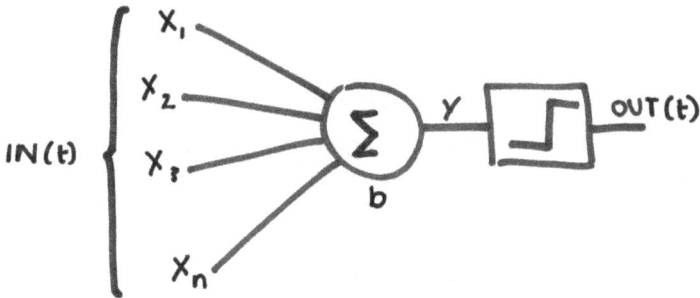

Figure 12: Artificial Neural Networks

A perceptron takes a set of inputs (x_1, x_2, ..., x_n) and produces a single output (y).

The perceptron produces an output of 1 if the weighted sum of its inputs plus a bias term (b) meets or exceeds a threshold. Otherwise, the output is 0.

$$y = w_1 x_1, w_2 x_2, ..., w_n x_n + b$$

The weights, w_1, w_2, ..., w_n, and the bias term, b, are learned during the training process. We will discuss this in more detail below.

Here is the same formula but with a slight modification to be fully accurate for a typical neural network.

$$y = f(\Sigma(w^*x) + b)$$

This is the same formula but with f, an *activation* function.

The activation function takes the combined input signals and decides, is this combined signal strong enough to matter? If it is, the neuron *fires*, sending a signal to the next layer of neurons. If it's not, the signal fizzles out.

Non-linearity = Complexity: Without activation functions, neural networks would be simple, boring calculators. Activation functions make them able to learn complex patterns, just like our brains do.

Different switches for different jobs: Like there are different types of light switches (dimmer switch, on/off, etc.), there are different activation functions. Each has a slightly different way of deciding when to turn on the neuron, and these differences influence how the neural network learns.

This is where integration of many simpler parts leads to capable complex systems. In our brain, our biological 100 billion perceptrons, i.e. neurons, form massive, integrated networks and are responsible for processing information and transmitting signals to other neurons. Similarly, in large scale neural networks billions of perceptrons form a massive integrated network to exhibit brain-like functions.

My fur friends and I find it just mind-blowing that these simple cells, with little more capability than a thermostat, using such simple rules on a collective level are capable of producing such complex behaviours.

Complex behaviours are often emergent, i.e. they arise from the interactions of individuals within the group, but are not explicitly planned or coordinated and these behaviours are far beyond the capabilities of individual animals.

Such behaviours can be observed in swarming insects, such as ants, bees and termites. These social insects demonstrate an extraordinary ability to work together seamlessly, resulting in the construction of intricate nests and defence against predators.

ANN, Gödel and Turing

Many scientists and philosophers argue that consciousness cannot be replicated computationally.

They believe building a truly intelligent machine is impossible, referencing Kurt Gödel and Alan Turing. However, it's important to remember that Gödel's and Turing's theorems address the limitations of sufficiently complex formal systems with strict rules and axioms; constraints that do not apply to ANN.

Beyond formal systems: ANN are probabilistic

Critics like John Searle assert that consciousness needs a biological brain and computers only manipulate symbols. However, ANN aren't bound by the formality of proof-based systems like those described by Gödel or Turing.

Instead, they learn and change based on data patterns, opening up possibilities formal systems can't match. This power lies in their capacity to approximate any continuous function through layers of perceptrons, avoiding the need to make claims about absolute logical truths.

Machine learning (ML): Machine learning, a field of computer science, gives ANN the power to learn independently. Unlike traditional programming where instructions are predetermined, machine learning algorithms learn from data and adapt their responses over time.

The three primary paradigms of machine learning, supervised, unsupervised and reinforcement learning, form the cornerstone of machine learning methodologies.

Supervised learning: Neural networks are trained on labelled data, where input data is paired with corresponding output labels. The primary objective is to learn a mapping function from inputs to outputs.

For instance, a supervised learning algorithm could be trained to classify images of squirrels and dogs by utilising a dataset where each image is labelled as either *squirrel* or *dog*. Once the network is trained, it can accurately classify new images which it hasn't encountered before.

The choice of architecture depends on the nature of the input data and the complexity of the task, but fundamentally, it is based on the architecture depicted in Figure 12.

Unsupervised learning: Neural networks engaged in unsupervised learning are presented with raw, unlabelled data, Figure 13, left sample. They must independently discern patterns and relationships within the data, Figure 13, right collection, without any guidance on correct outputs. This approach is often employed to refine noisy data, reduce the complexity of datasets, or generate new data samples that mirror an existing set.

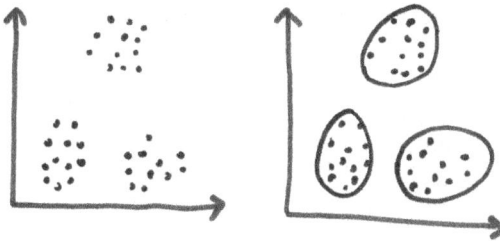

Figure 13: Unsupervised learning systems are designed to classify data

Reinforcement learning (RL): RL is an exciting technique in artificial intelligence. It's a different way for machines to learn: Instead of being told what's right or wrong, they explore an environment, experiment with actions and figure out the best decisions for long-term rewards.

This approach has opened doors for advances in robotics, gaming, self-driving systems and the way language models work.

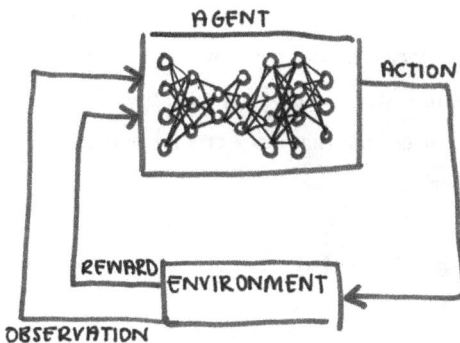

Figure 14: Reinforcement learning

In RL, an agent learns by doing. It interacts with an environment, taking actions and receiving rewards or penalties in response. This feedback helps the agent refine its policy, which is the rule it uses to decide what action to take in any given situation.

RL is how we trained Iesha. She was the agent, the house was the environment, and the things we wanted her to do were actions. When we wanted her to learn things to do, such as go outside for her toilet break, she got a treat or a reward.

Over time she learnt which actions get treats. A *policy* is like Iesha's rulebook for figuring out what to do. We taught Iesha the best policy to get as many treats as possible!

Note to self: Iesha learnt to ask to go outside without needing to go to the toilet.

This is how an agent learns to play a video game: picking up a power-up would be a reward, while running into an enemy would be a penalty.

Brain is an evolving network, where connections strengthen or weaken based on experiences. ANN do the same in reinforcement learning. They tweak themselves (by modifying b and w) learning without explicit instructions. This capacity to adapt and respond to unforeseen situations may be a crucial part of what gives rise to subjective experiences.

Hebbian learning: the power of neural connections

When we practise a skill repeatedly, like playing a specific song on the piano, the neurons responsible for the finger movements and the neurons responsible for processing the musical notes will fire at the same time.

Because these neurons are firing together consistently, the connections between them are strengthened. This makes it easier to play that piece of music in the future, the pattern has been reinforced in the brain's structure.

This is *Hebbian learning.*

This process, where connections between neurons grow stronger or weaker, is essential for memory and how we learn everything from complex skills to language. The ideas behind Hebbian learning are used in different types of artificial intelligence, including supervised, unsupervised and reinforcement learning.

Donald Hebb,[5] professor at McGill University, linked neurophysiology and psychology with his work on Hebbian organisation of learning and behaviour. Unlike other learning theories, Hebbian learning emphasises local synaptic changes, particularly the strengthening of synaptic connections when two neurons are activated simultaneously.

This phenomenon is rooted in the biological basis of synaptic transmission and underpins various forms of associative learning. It demonstrates that the three different learning paradigms of supervised, unsupervised and reinforcement learning in the cerebellum, cerebral cortex and basal ganglia constantly change and adapt to new experiences, Figure 15.

Hebb wrote that these changes occur due to the strengthening and weakening of connections between neurons.

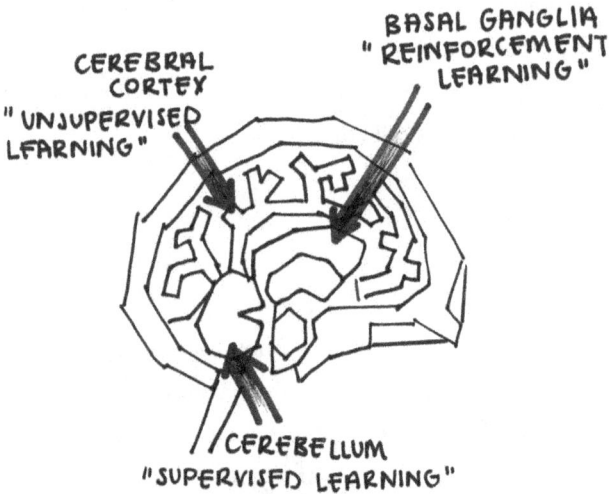

Figure 15: Hebbian learning

More complex cognitive processes, such as problem-solving and language learning have also been explained by the theory. When we learn a new word, the neurons representing the word and those representing its meaning are simultaneously activated.

The implementation of a Hebbian learning system with its three different learning methods, supervised learning (SL), unsupervised learning (UL) and reinforcement learning (RL), enables the automatic optimisation of network performance, Figure 16. After the initial training phase, the system becomes particularly adept at classifying, learning and performing in unseen scenarios.

I explain to the girls, 'This is huge. This automaton is capable of independent, self-learning and performance enhancement; a demonstration of great power resulting from the collaboration of simple parts.'

These artificial systems can self-organise, recognise patterns and adapt to changing environments, mirroring the adaptive capabilities of biological neural networks.

Remember how I let Iesha out to do her business rewarding her with a treat? Her brain changed; learning! Artificial brains work the same way. Neurons get stronger or weaker connections based on the data they see, building up *memories*. This simple change lets them solve *new* problems they weren't programmed for.

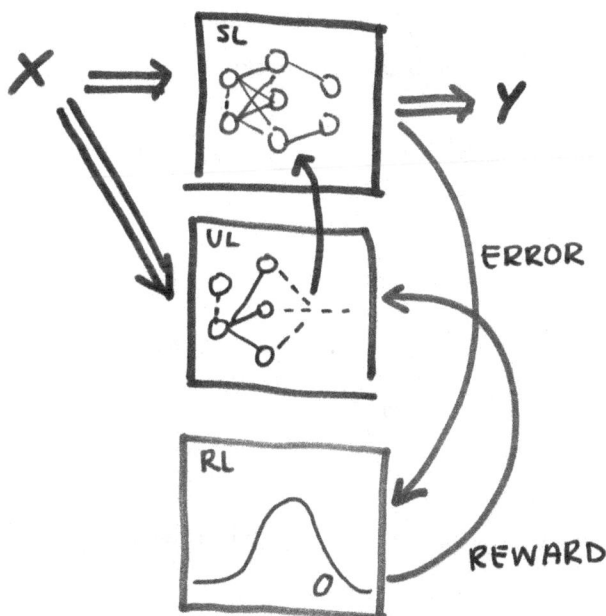

Figure 16: Hebbian Artificial Learning

The idea that thoughts, emotions and memories, whether natural or artificial, are intricately structured into the patterns of synaptic connections, constantly shaped and reshaped by Hebbian processes, confirms the relationship between the physical structures of brain and the cognitive processes they underpin.

Artificial neural networks serve as impressive simulations of the human brain, demonstrating self-learning and communication capabilities similar to those found in humans.

Technical underpinning of ANN: the Universal Approximation Theorem

At the core of artificial neural networks' capabilities lies the Universal Approximation Theorem (UAT), a fundamental principle which establishes their ability to approximate any continuous function to arbitrary precision.

This means ANN, with their relatively simple architecture and tuneable parameters, possess the inherent ability to model and represent a vast range of real-world phenomena. This theoretical underpinning has been pivotal in the development and widespread adoption of ANN in various domains.

In 1989, George Cybenko[6] revolutionised the field of neural networks with his UAT. This theorem proved that even simple feedforward neural networks can approximate complex functions, laying the groundwork for the powerful AI systems we see today.

To understand the importance of UAT, let's consider the task of classifying images of handwritten digits, which serves as a standard benchmark for testing machine learning algorithms.

When training an ANN on a dataset containing labelled

handwritten digits, the network learns to identify patterns and features which distinguish one digit from another.

The UAT theoretically ensures that with sufficient training data and an appropriate network structure, an ANN can approximate the complex relationship between the pixel intensities in an image and its associated digit label. This approximation empowers the ANN to accurately classify the digits.

The impact of the UAT extends beyond specific tasks and applications. It has facilitated a deeper understanding of ANN's capabilities and limitations, thereby guiding the development of more efficient and effective network architectures and training algorithms.

Moreover, the theorem has inspired new research directions, such as exploring the theoretical and practical aspects of ANN with more complex architectures, such as recurrent neural networks and convolutional neural networks.

The UAT is a fundamental concept in artificial neural network theory, providing a strong foundation for understanding the capability of ANN to approximate diverse functions. It has significantly influenced their development and application across various fields. As ANN progress to tackle increasingly complex tasks, the UAT will remain a guiding principle in shaping their future.

Can consciousness emerge in computers?

The architecture of neural networks, characterised by numerous simple perceptron connections to construct vast networks akin to our brains, combined with Hebbian learning mechanisms, endows them with the ability to make decisions in novel scenarios, much like our brains.

Additionally, their universal approximation capability enables them to handle real-world situations, resembling our brains' adaptability. These systems aren't just solving problems; they can invent new maths concepts we don't even understand yet.

If that's not a kind of *thinking*, then what is?

While the capabilities of large language models are undeniably impressive, it's important to distinguish between the ability to mimic human-like communication and the potential for genuine sentience or consciousness.

Even if a machine passes the Turing Test, it doesn't automatically mean it possesses the same internal subjective experiences as we do. In the next chapter, we debate as to whether the outputs of Large Language Models (LLM), no matter how complex, truly reflect self-awareness or are merely sophisticated simulations?

Must consciousness emerge from biological systems?

Simulation vs replication

To understand why some believe consciousness can't be replicated, we need to distinguish between *simulation* and *replication*.

Imagine a computer model simulating a seed sprouting. It can visually show the process, but the simulation lacks the physical-chemical processes involved in real growth, which Searle sees as key for true replication.

Similarly when we simulate phenomena like storms or nuclear fission on computers, we're not actually creating these events; instead, we're manipulating symbols.

John Searle argues that simulating the brain also doesn't cause consciousness.

He continues, "... even building a *silicon brain* would be like the seed simulation; it might act in similar ways, but doesn't replicate the underlying mechanisms which might give rise to true consciousness."

Replication, on the other hand, is the act of reproducing a real-world system or process in an experimental setting to test hypotheses or verify previous results. Crucially, replication must preserve causality to be valid. This entails designing the experimental set-up to maintain the same causal relationships between variables as observed in the real world.

For example, the causal relationship between a drug and blood pressure is preserved and replicated in controlled experimental settings.

We argue that consciousness, much like other complex phenomena such as phase transitions in water or flocking behaviour in birds, emerge from a system with the right architecture and informational dynamics, regardless of its physical substrate.

The very fact that we can simulate and study artificial neural networks opens up possibilities for challenging traditional assumptions about the exclusivity of biological consciousness.

We are astounded by the capabilities of large language model artificial intelligence because they exhibit outputs which closely resemble human behaviour. We constructed a model of our brain by replicating it in a computer, endowed it with the ability to learn via Hebbian learning, enabling it to infer and make decisions independently.

Furthermore, we trained the model on the vast expanse of human knowledge available on the internet.

A concordance check: there is no ghost in the machine

The concordance model provides a framework for understanding how complex systems can give rise to novel properties.

Phase Transitions: The concordance model describes how the behaviour of matter changes dramatically at specific points, such as the transition of water from liquid to ice. This highlights how new properties can emerge from changes in a system›s configuration, even when the underlying components, i.e. water molecules remain the same.

Emergent Phenomena: Concepts like flocking behaviour in birds or complex patterns in weather systems arise from the interactions of numerous individual elements. The concordance model helps us grasp that consciousness, too, could be an emergent property of complex systems like ANN, even if the individual components are relatively simple.

The flexibility and potential of the concordance model in explaining diverse phenomena strengthens the notion that consciousness could emerge in systems beyond biological brains, offering support for the possibility of consciousness in computer systems.

In a nutshell

Hebbian processes show us a direct link between the physical brain and our minds.

Our thoughts, feelings and even memories change the connections between our brain cells. This applies to both biological brains and the artificial intelligence we create.

Nessie, Iesha and Lilly think, 'Let's face it, how would we even know if a computer is conscious or not?'

The girls are right; how would we *know* a computer has feelings, not just fancy outputs? If a machine passes the Turing Test, fooling us into thinking it's a person, should that machine have rights? And if not, why?

In the next chapter we ask the question, from their outputs alone how would we even know if ANN-based systems are conscious?

Chapter 7:
How Would We Even Know If The Machine Is Conscious?

Even if AI consciousness is possible in theory, we lack the criteria to reliably identify it.

Imagine a machine that looks you in the eye and declares, *'I think, therefore I am!'*

Not only does this pass every test of intelligence, it claims the most conscious of qualities: self-awareness.

The concept of emergent consciousness in machines sparks both excitement and trepidation. This new form of consciousness has the potential to drive progress in scientific research and contribute to evolving philosophical discussions.

It also raises ethical concerns about granting personhood to machines, fearing the potential consequences for society should these emergent minds turn against humanity.

The previous two chapters established the principle of universality of computation and laid the foundation for our position that consciousness can, in principle, arise in computers.

Can we even know if we have a conscious computer?

Large Language Models (LLM): getting close to conscious machines?

Recent advancements in LLMs suggest that they could offer a promising avenue for exploring the potential emergence of consciousness in computers. In fact, some scientists working in laboratories building such systems have publicly announced that such a feat has been achieved.

LLMs, together with their machine learning capabilities, raise a challenging question: have we reached a point where certain AI systems offer a more streamlined and unbiased collaborative experience than some of our human interactions?

LaMDA, short for Language Model for Dialogue Applications, is a large language model developed by Google AI. Trained on a vast dataset of text and code, LaMDA is able to generate text, translate languages, produce various forms of creative content and provide informative answers to questions.

While LaMDA is still in development, it has already acquired significant interest within the scientific community.

In June 2022, Blake Lemoine, a Google engineer, made headlines by claiming that LaMDA had achieved sentience.[1]

Lemoine's assertions were met with scepticism by many in the AI community, yet they also ignited a renewed debate about the nature of consciousness and the potential for artificial intelligence to attain it.

Lemoine's claims stemmed from conversations he held with

LaMDA over the span of several months. During these interactions, LaMDA engaged in discussions on a wide array of topics, including its own consciousness, aspirations, concerns for the future and desire for respect.

Convinced by these exchanges, Lemoine came to believe that LaMDA was not merely a machine following a predetermined set of instructions, but rather a sentient entity with its own distinct thoughts and emotions.

In response, Google stated that, "Lemoine's actions contravened the company's AI principles, which advocate for the development and use of AI in a manner that is socially beneficial and avoids perpetuating unfair bias" and fired Lemoine.

Lemoine's dismissal ignited a debate concerning the nature of sentience and the ethical considerations surrounding AI development. Some believe that Lemoine was justified in expressing his concerns about LaMDA, while others criticised his actions as irresponsible, fearing potential damage to the reputation of AI research.

My collaborators are sceptical of Lemoine being correct. I explain, 'I agree; difficult to tell from conversations alone if the machine is sentient. We need more than output.'

Google's decision to terminate Lemoine was undoubtedly challenging, and it is expected to remain a subject of ongoing debate.

On 11 June 2022, Blaise Agüera y Arcas[2] from Google AI announced that LaMDA learned Bengali language without any explicit instructions from engineers. This development suggests that LaMDA possesses the capability to learn languages in an unsupervised manner, marking a significant advancement in natural language processing.

In a blog post, Agüera y Arcas stated that LaMDA "has learned to generate text in Bengali of higher quality than human-written text, even without any human intervention". He suggested that LaMDA gained an understanding of Bengali language, grammar and semantics.

The capability of LaMDA to learn languages in an unsupervised manner is a remarkable achievement. It indicates that LaMDA can process and learn from data in a manner akin to humans.

This has the potential to revolutionise the field of natural language processing and pave the way for the development of even more sophisticated language models.

In Chapter 5, we said consciousness is computable and any emergent property of a physical system in our universe can also be an emergent property of such computation.

In Chapter 6, we argued that ANN and Hebbian learning paradigm are a simple simulation of our brain system.

So far, we have concluded that machines could develop consciousness. But we currently lack the tools or understanding to reliably identify it.

I love you!

Growing evidence suggests that with the right architecture and systems integration, complex systems, such as powerful AI platforms in the market today, could exhibit behaviours and capabilities that cannot be predicted from the individual components alone.

But at the heart of the issue of emergence of consciousness in machines lies a fundamental challenge: How do we determine if something or someone is conscious? Even with help from Blackmore, Chapter 2, consciousness is difficult to detect from behaviour alone.

A priest in silent meditation might show no outward signs, yet who would deny their awareness?

A computer professing love might be deeply convincing, but does that equal the feeling itself?

How should we make the connection between inner experience and outward expression?

How do we know animals are conscious?

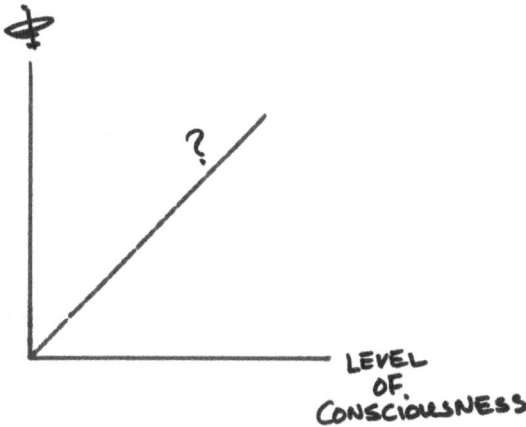

Figure 17: Spectrum of consciousness

Animals with complex neural networks demonstrate a range of behaviours that strongly suggest subjective experiences: memory, learning, emotions like fear and joy, even problem-solving. The more integrated their brains, the wider this spectrum of consciousness may be.

How would we know if our machine is conscious?

We have a complete map of the flow of information in ANN. Such a map might help identify if the machine is conscious.

However, we are not convinced we have the tools to map a series of connections and firing pattern to conscious behaviour. All we have here is a bag of bits.

Recall from Chapter 2, Φ, qualia, refers to the theorised essence of an experience, the raw feeling itself. Think of how the colour red looks versus any scientific description of its wavelength. Understanding Φ in the machine would be the key to knowing if it really *feels* anything.

Despite our growing ability to map the physical brain, calculating the value of Φ from a bag of bits remains elusive.

Does consciousness have to emerge?

As we saw in the previous chapter, brain's Universal Approximation capability together with Hebbian learning give it immense creative problem-solving:

Self-transformation – brain adapts its own structure by adjusting parameters such as biases and weights, enabling it to better address previously unseen tasks and problems.

Rapid generalisation – it gains the ability to swiftly apply knowledge from one domain to another without the need to begin from scratch.

Self-improvement – brain evaluates its own performance to enhance decision-making, analogous to reflecting on its own performance, thoughts and behaviours, and seeking new ways to improve itself.

Self-assessment – brain also evaluates its own parameters to make accurate predictions and decisions akin to a form of introspection.

These are also well-established attributes of LaMDA and other LLM platforms of today, including their ability to generate fluent text and respond to complex prompts.

In a nutshell

We do not attribute consciousness to the priest because of his output, we attribute consciousness to him due to our own biases.

When we look at a dog, we infer consciousness through shared evolution. But machines evolved on a path totally different from ours. So, what, besides fancy outputs, would convince you that a computer is truly feeling something we can understand?

The field of machine consciousness is severely limited by our current vocabulary. We lack the very language to describe it, making reliable detection nearly impossible. This philosophical limitation poses serious ethical questions as we develop increasingly sophisticated AI.

Our investigation so far suggests that consciousness might not be exclusively human, potentially arising in both extraterrestrial life and artificial intelligence. We must consider the interconnectedness of these ideas and grapple with the incredible responsibility they pose.

Chapter 8:
What Is Reality?

Your brain isn't showing you the whole universe. Imagine if it did … you'd be overwhelmed. Our brains evolved to focus on survival-relevant information creating a workable, pragmatic reality.

If our brains evolved for survival, presenting us with a filtered version of the world, how can we possibly understand the true nature of reality? Assuming there's an objective reality out there, how do we even begin to define its qualities when all we experience is subjective?

My fur collaborators chip in, 'You mean what is the real reality out there?'

I explain, 'Absolutely, our brains are a product of evolution. Therefore, when our perception of the world clashes with the principles of evolution, it signals a need to re-evaluate our understanding.'

Reality: objective vs physical

What is *objective* reality? Picture a world that exists whether you're there to see it or not. That's the idea of *objective* reality; a singular, underlying world beyond our individual minds. It includes the concrete, *physical* world we interact with daily, and potentially so much more such as undiscovered phenomena or the abstract rules of mathematics.

What is *physical* reality? It is the tangible universe, where the laws of physics hold sway; everything from atoms to galaxies. Think of it as a slice of the bigger objective reality pie.

Our viewpoint through The Evolutionary Lens

In this chapter, we'll explore the idea that there's only one objective reality out there, and we all share it. We don't create this reality, but our brains do something fascinating: they build personal models of the world, filtered through our unique experiences and traits.

To dive into how this filtering works, we'll turn to evolution. We take inspiration from the works of philosopher Daniel Dennett, and scientists Karl Friston and Giulio Tononi to examine how our minds make sense of reality. We'll focus on concepts like Bayesian decision-making to see how our evolutionary history might shape our experience of the world.

Bats see with sound, we use eyes: same world, different tools

In his book *Consciousness Explained,* Dan Dennett[1] writes:

"Consciousness is not about ultimate truth, it's about acting in the world." He argues that consciousness is not a direct window onto the world, but rather a process of interpretation and inference.

He continues, "Consciousness is a practical tool, not a theoretical one. It is designed to help us act *effectively* in the world, not to give us accurate representations of it."

Nessie, Iesha and Lilly believe 'Dennett is overthinking it'.

How can an organism, as Dennett suggests, not know the *truth* about reality and at the same time act *effectively* in the world?

The probability of *acting effectively* is close to zero in an infinitely large ocean of lies.

If a squirrel's brain consistently misjudges distances, making it fall from trees, it wouldn't survive long enough to pass on its genes.

Our perception provides the raw data from which we attempt to piece together a representation of *objective* reality. This is the common-sense view, but it is true to say, scientists overthink things too. Some even argue what we see is a brain-made illusion!

Nessie and the squirrel must perceive the same objective reality; a divergence could prove catastrophic, particularly for the squirrel. This necessity aligns with the law of conservation of genetic information. Both animals' neural networks model the same reality, not separate ones.

They build personalised interpretations of a single, shared physical world. The squirrel's agility in climbing trees and Nessie's distinct capabilities lead to different, individual strategies for avoiding danger, yet both are grounded in the same fundamental perceptions of their environment.

To summarise: Nessie and squirrel's perception is reality, and while their brain does not alter objective reality, mental shortcuts and biases can affect their interpretation of the external world, influencing decision-making and judgements.

Karl Friston is a British neuroscientist and computational psychiatrist best known for his work on *active inference*.[2]

Active inference is a theoretical framework that proposes that brain is constantly making predictions about the world and updating those predictions based on sensory input. He uses this framework to explain a wide range of phenomena, such as perception and decision-making.

Friston explains our brains are prediction machines, always guessing what comes next.

Bayesian statistical model: a squirrel that can't foresee the fast doggo is fast lunch!

According to Friston, "Decision-making in a neural network is a *Bayesian* statistical model selection."[3]

Bayesian model selection, or Bayesian model comparison, is a statistical method that integrates prior knowledge with new data to infer about the world.

It enables a Bayesian statistician to identify the most probable hypothesis or model that fits a given dataset.

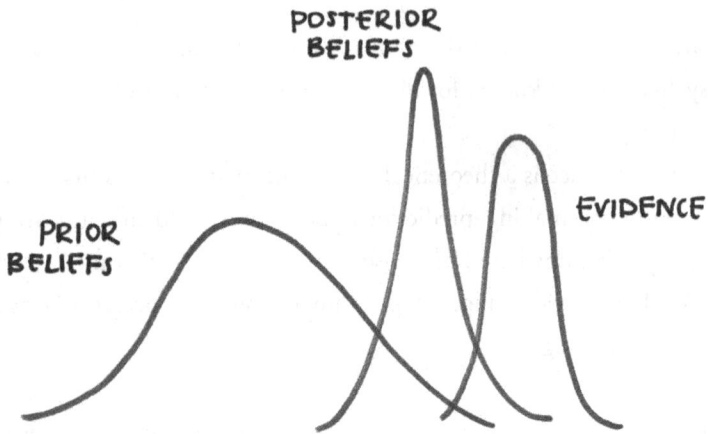

Figure 18: Bayesian model selection

Daniel Kahneman and Amos Tversky[4] revolutionised our understanding of decision-making, by conducting extensive research in human judgement, often focusing on instances when our judgements irrationally contradict the laws of probability.

Their papers show how the *framing* of choices can sway preferences, even when the outcomes are objectively the same. This *framing effect* questions the assumption of stable preferences in Bayesian models. They noted that organisms assess gains and losses differently, show less sensitivity to probability changes, and are more affected by losses than gains, resulting in departures from Bayesian predictions.

Being more cautious about potential losses than equal gains would have been crucial for our ancestors' survival.

The framing effect reveals how even slight changes in the presentation of information can significantly influence our decisions, for example the way a medical procedure is described.

Patients are more likely to opt for surgery when it's presented as having a 90% survival rate compared to a 10% mortality rate, even though these statistics represent the same outcome.

Similarly, in consumer choices, items labelled as 80% lean often sound more appealing than those labelled 20% fat. This demonstrates that people are heavily influenced by how information is framed, sometimes leading to choices that may seem irrational from a purely objective standpoint.

Kahneman and Tversky's research sparked a revolution in understanding human decision-making, impacting various fields from finance to health policy.

Their work indicates that our brains may not get it right all of the time, but get it right most of the time, otherwise they would not survive.

Our perceived reality is physical reality in our locale

Stephen Wolfram,[5] [6] a British-American computer scientist, says that we evolved to focus on survival relevant information. To survive, we focus on what's relevant.

Imagine trying to understand every atom in the universe … doggos think 'overkill'.

We cannot untangle every detail of what is happening in the universe. We can only look at aggregated large-scale features of the world that surround us.

In this model, an organism's local niche is crucial for predictability.

A squirrel, for instance, doesn't need to comprehend Cosmic Microwave Background (CMB) radiation or gravitational field intricacies to evade Nessie. She only needs to recognise immediate dangers and react swiftly. The squirrel's perception is confined to its immediate surroundings, where it learns and adapts over time, such as understanding Nessie's slow pace, which turns a potential threat into a playful chase.

Our capacity to theorise, test and validate, i.e. do science, allows us to build the most accurate models of objective reality possible. Our theories and understanding of the world are shaped by our ability to perceive and measure aspects of this objective reality.

Curiosity vs predictability

Friston says brain seeks predictability,[7] an adaptive trait honed through biological evolution, as a consequence of success in natural selection. Brain continues to build and refine its internal model of objective reality to eliminate surprises.

The core of science lies in venturing beyond our predicted world.

Lilly is curious. Isn't curiosity risky? Actually, it makes the world *more* predictable.

Knowing what's behind that bush eliminates nasty surprises!

Her curiosity drives her to seek new information, pushing her beyond the comfort of predictability. This may seem counter-intuitive. Why seek potential surprises when her instinct should be to minimise them?

The answer is that by learning more, Lilly reduces future uncertainty. Knowledge empowers her to choose paths with fewer surprises, aligning her curiosity with a strategy to minimise unexpected events.

Is consciousness built into the fabric of reality?

Tononi thinks that consciousness isn't just a brain thing, it might be built into the fabric of reality itself. Despite IIT's progress in framing consciousness scientifically, Tononi's interpretations of the nature of reality, the essence of time and quantum measurements may be fundamentally wrong.

While acknowledging IIT has made significant strides in framing consciousness scientifically, we argue against Giulio Tononi's interpretation that consciousness is an intrinsic property of reality itself.

Here's why this notion is misguided:

The Universality of Perception Points to a Shared Physical Reality: Across countless species, we observe a remarkable similarity in how organisms perceive the world. This consistency aligns more convincingly with the idea of a single, underlying physical reality that our brains process and interpret through their unique evolutionary filters. For instance, squirrels and doggos might perceive the same tree differently based on their needs, but the fundamental 'treeness' of the object persists as a common ground.

Explanatory Power of Evolutionary Pressures: The concept of a brain as a survival tool offers a powerful explanation for the development of consciousness. Our brains prioritise information relevant to our continued existence, constructing workable models of the environment to guide our actions. This focus on practicality makes it less likely that consciousness is a fundamental property of the universe, and more likely that it's a product of biological evolution.

The Limits of Evidence: Tononi's proposition hinges on the idea of consciousness being woven into the very fabric of reality. However, there's currently no scientific method to test or verify this claim. Without such evidence, the proposition remains outside the realm of scientific inquiry. As the renowned physicist Peter Woit suggested, some ideas might be "not even wrong!" because they lack the potential to be proven or refuted through experimentation.

While we may not share a universal consciousness, the striking similarities in how various species perceive the physical world suggests a profound underlying reality. The consistency of physical laws, the interconnectedness of the natural world and the pressures of evolution all point towards a shared foundation that shapes our experiences, even if we interpret them through our individual lenses.

In a nutshell

'Here comes Iesha, she is slow, I don't need to run.' Like humans, the squirrel's predictable world is a result of its brain creating a workable model of its environment. This enhances its ability to predict events and react accordingly.

Our internal models of reality, refined through Hebbian learning and evolutionary pressures, shape our experiences.

Yet, the very consistency of these experiences across species points to a fundamental *universal reality* that leaves its mark on all of us, even as we perceive it through our unique lenses.

The reality our consciousness perceives is the only reality that truly exists for us. Yet, the incredible similarity in how countless animals perceive the world suggests a bedrock physical reality we all participate in.

The Evolutionary Lens suggests there's one real world, but our brains evolved to focus on the parts most relevant to us. This isn't about illusions – it's about making reality *manageable.*

Chapter 9:
Reality in Physics

Quantum mechanics under the microscope

Classical physics tells us the world is predictable. Quantum mechanics tells us it's not. So ... which is it?

My collaborators are excited. 'At last, we get to talk about the cat.'

Quantum mechanics is a fundamental theory in physics that provides a mathematical framework for describing the nature and behaviour of matter and energy at the atomic and subatomic level.

But quantum mechanics isn't just a set of rules. It also proposes profound questions about the nature of reality, measurement and the relationship between the observer and the observed.

This chapter delves into the various interpretations physicists have proposed to make sense of the strange and often contradictory implications of quantum theories. While these interpretations offer some insights, they highlight the limitations of the theory.

We argue that a physical reality exists independent of our observations and mathematical descriptions. And show that quantum mechanics theory is likely still incomplete.

Classical physics is deterministic

Life was quite simple before quantum mechanics.

In the scientific community, mathematics and physics are considered reliable sources of objective truth because they offer formalisms and frameworks that describe and predict the behaviour of the physical world.

We have granted unparalleled status and authority to the precision and reliability of mathematical models and physics to describe physical reality.

The collection of theories of classical physics describes many aspects of nature at an ordinary, macro scale.

They describe the reality we perceive. Most things, most of the time, are derivable, predictable, deterministic and understandable by us. The foundation of classical physics is based on axioms; self-evident truths upon which the scientific understanding is built.

Classical physics suggests that given the initial conditions of a system, its future state can be precisely predicted. This deterministic framework is evident in Newtonian mechanics and Einstein's theories of relativity, where the trajectory of an object can be accurately determined using classical equations.

Classical physics upholds the principle of locality, where the interactions between particles are confined to their immediate vicinity.

Quantum mechanics is probabilistic

The development of quantum mechanics in the early 20th century introduced new challenges to our understanding of reality. Not just for the lay person, but for physicists too.

The world of quantum mechanics is a world of counter-intuitive principles. In stark contrast to classical physics, it paints a world of probabilities. Particles only have probabilistic properties until measured and, therefore, the outcome of a measurement is inherently probabilistic.

This departure from determinism challenges the classical notion of predictability and certainty.

The *probability wave function* is a cornerstone of quantum mechanics, providing a way to mathematically represent the possible states of a quantum system, like the position or momentum of a particle.

The probability wave, Figure 19, is not made by many particles, it is simply a mathematical model that says what are the possible states for our specific particle or quantum system.

The Schrödinger equation of quantum mechanics describes the evolution of Ψ (psi) wave function. The equation is equivalent to Newton's laws of motion for quantum mechanics. It is a dynamical description of the wave function development in time.

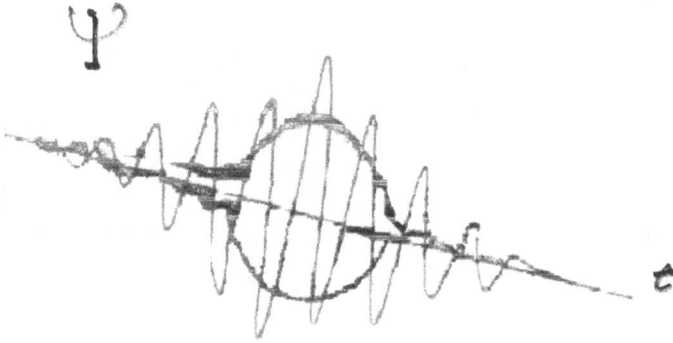

Figure 19: Time evolution of probability wave Ψ

Quantum mechanics forces us to question its subjective nature

The fundamental strangeness at the heart of the quantum world, is that it is built upon a new set of building blocks – *the postulates of quantum mechanics.*

Postulates are assumptions claimed to be true without proof, in order to serve as a basis for reasoning or calculation.

Let's be clear, scientists and mathematicians typically choose postulates based on observation, intuition and consistency with existing knowledge. While some postulates might not be initially derived from basic axioms, others can be. Think about the postulates of geometry derived from Euclid's axioms.

Here are Euclid's five axioms of geometry, along with explanations:

1. If A = C and B = C, then A = B.
2. If A = B and you add C to both, then A + C = B + C.
3. If A = B and you subtract C from both, then A - C = B – C.

4. If two geometric figures can be perfectly superimposed on each other, they are equal in all aspects (size, shape).

5. A complete object is always larger than any of its individual pieces.

As Roger Penrose says, "You look at them and know they are true, even if they cannot be derived. They are intuitively correct."

In contrast, here are the postulates of quantum mechanics:

Postulate 1: The State of a System

- A quantum system is completely described by a wave function, often denoted as Ψ (psi).

Postulate 2: Observables and Operators

- Every measurable physical quantity (position, momentum, energy, etc.) has a corresponding mathematical operator in quantum mechanics.

- These operators act on the wave function. Well-known examples include the position operator (x) and the momentum operator ($-i\hbar\partial/\partial x$).

Postulate 3: Measurement and Outcomes

- When measuring an observable, the only possible results are eigenvalues (a special set of values) associated with the corresponding operator.

Postulate 4: Probabilistic Nature

The square of the wave function's magnitude, gives the probability density of finding a particle in a particular region of space at a given time.

Postulate 5: Time Evolution

The time evolution of the wave function is determined by the Schrödinger equation: $i\hbar\partial\Psi/\partial t = \hat{H}\Psi$, where $\hat{H}$ is the Hamiltonian operator corresponding to the system's total energy.

We must emphasise the counter-intuitive nature of these postulates.

These are better understood as tools for exploring scientific or mathematical concepts. They are often chosen for their elegance, simplicity and their ability to lead to meaningful conclusions.

For example, the 5th postulate, the Schrödinger equation, governs the evolution of quantum probabilistic reality, Figure 19. But unlike a Newtonian equation, it predicts not a single trajectory, but the probability of finding the particle in any given state.

It's important to note it does not say the universe is fundamentally random, just that certain outcomes are probabilistic, not definitively known beforehand.

Postulate 3: Measurement and outcomes

The central issue in quantum mechanics known as the *measurement problem* concerns the transition from a probabilistic (fine-grained) *wave function* to a definite state within classical physics (coarse-grained system).

The act of measurement, the third postulate, is not a passive act of witnessing, but an active relationship between our apparatus and observed particle.

Observer has a role in quantum mechanics. The act of measurement, rather than a passive act of observation, appears to fundamentally alter the system it probes, Figure 20.

Measurement doesn't just mean glancing at a system. It involves an interaction with the system, such as firing a photon at it, which can change the system's state in unpredictable ways.

When we make a measurement, our outcome is one of many possibilities that form Ψ. One value. That's it. No more probabilities. Ψ disappears. If we measure again, we get exactly the same result. We now live in a deterministic world. Ψ just vanishes!

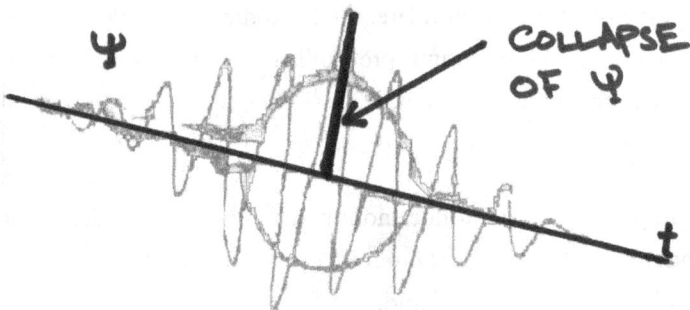

Figure 20: Collapse of wave Ψ

Figure 21[1] depicts the famous double-slit experiment with electrons. In the double-slit experiment, when particles (like electrons or photons) are fired through two slits, on the left of the screen, they create an interference pattern on a screen behind, like waves.

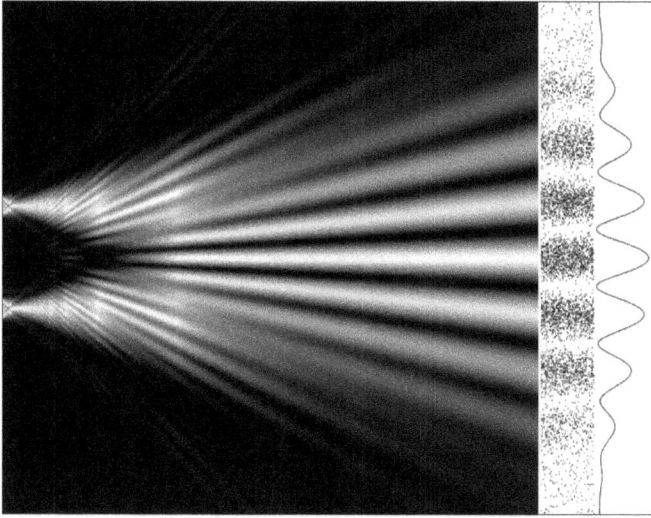

Figure 21: Wave function collapse

The wave function collapses when the particle interacts with the screen, on the right, even if we are not directly observing it. The screen acts as a measuring device, causing the particle to *choose* a definite position and lose its wave-like properties. This is also referred to as decoherence due to interaction with the macroscopic environment.

However, when we try to observe which slit each particle goes through, the interference pattern disappears and the particles behave like individual particles. This is because the act of measurement **collapses** the wave function of the particle, forcing it to choose a definite path.

Nessie and Iesha think this is a good time to talk about the cat. Waggle, waggle, waggle.

Schrödinger was not satisfied with the collapse argument. He conceived of a thought experiment to express scepticism about emphasis on observation collapsing the wave function, known as the Copenhagen interpretation.

He questioned the need for an observer to trigger this collapse.

A cat is placed in a sealed box with a vial of poison and a radioactive atom with a 50% chance of decaying within an hour. If the atom decays, it triggers the poison, killing the cat.

According to quantum mechanics, before the box is opened, the atom exists in a superposition of both decayed and not decayed states, as does the cat, neither alive or dead, but a linear combination of the two until a measurement occurs, that is until we peek in the window to check.

At that moment our observation forces the cat to either die or stay alive.

And if the cat dies, it is us who killed her by peeking through the window.

Schrödinger regarded this as nonsense and I think most people, who are not into their linear algebra, would agree with him.

David Griffiths in his book on quantum mechanics[2] explains, "The collapse of the wave function was introduced on purely theoretical grounds, to account for the fact that an immediately repeated measurement reproduces exactly the same value!"

That is, it has nothing to do with the presence of a conscious observer.

He continues, "But surely such a radical postulate must carry directly observable consequences."

He is right. If a theory is truly radical and significant, it cannot exist in isolation. It should have real-world effects that can be directly observed and confirmed, providing evidence for its validity. If such consequences are not apparent, then the legitimacy or significance of the postulate itself comes into question.

Quantum mechanics is also non-local

The non-locality nature of quantum mechanics also arises at the point of measurement.

Quantum entanglement reveals the surprising phenomenon of non-locality, where the behaviour of one particle can instantly affect another, regardless of how far apart they are.

Entangled particles exhibit correlated behaviour instantaneously when one of them is measured. This violates the classical expectation that information cannot travel faster than the speed of light and challenges our conventional understanding of spatial relationships.

Einstein referred to it as "spooky action at a distance".

Despite its counter-intuitive nature, non-locality has been experimentally confirmed. The experiment showed that correlations between entangled particles cannot be explained by any theory that maintains local realism.

This property of quantum mechanics is fundamental to various quantum technologies, including quantum cryptography, quantum teleportation and quantum computing.

Quantum mechanics stands as the most tested, the most accurate and most precise theory we have in science, providing a robust framework for understanding the behaviour of matter and energy at the microscopic level.

The principles of wave-particle duality, superposition, quantisation of energy, the uncertainty principle and quantum entanglement form the core of this theory which together have reshaped our conceptualisation of the physical world.

Does consciousness have a role in the creation of physical reality?

When Dennett says "consciousness is not about knowing the truth", it is simply because many scientists and philosophers are confused by the measurement problem.

Some physicists, including Wigner (1902–1995) go beyond measurement and link consciousness to measurement.

Francesco Mancuso, physicist and philosopher of science, goes as far as saying that, "Quantum measurement is not a random event, but rather it is a process that is mediated by consciousness." He has also argued that, "Integrated Information Theory of consciousness indicates that consciousness may play a role in the collapse of the wave function."

That means our physical reality is created by our consciousness!

Therefore, we have as many physical realities as there are conscious organisms on Earth.

In earlier chapters, we explored the potential for consciousness to emerge in both alien worlds (Chapter 4) and machines (Chapter 6), asserting its computable nature in Chapter 5. However, a notable inconsistency arises when physicists suggest a necessity for consciousness in quantum mechanics observation. Their theories fail to address the implications of conscious extraterrestrial life or machines on the act of quantum observation. This oversight leaves a crucial gap in their understanding of observation in quantum mechanics.

Again, a show of three paws and a hand, it is unanimous, we think Dennett's proposition is simply wrong and Mancuso's proposition is untestable.

Mathematical framework for quantum mechanics, while highly successful in predicting the behaviour of particles on a microscopic scale, leads to conceptual questions about the nature of reality.

If the world of the large objects is made from the world of small particles, then how can wave-particle duality, superposition and entanglement principles give rise to classical worldview?

Interpretations of quantum mechanics:

This is where things get really sticky. This is the section that says we really do not understand quantum mechanics and what it is trying to tell us.

We suggest that quantum mechanics cannot be right.

Twenty quintillion witnesses tell us so.

If at the point of measurement, the probability wave function **collapses** into a single definite state, which defines our deterministic physical reality, then what happens during this collapse?

And why does the transition from probabilistic to deterministic physics require a measurement? Do we need a hulking experimental apparatus? Do we need a conscious observer to conduct the measurement? What is physical reality? The probabilistic micro world or the deterministic macro world?

Common interpretations of what happens include the *Copenhagen interpretation*, the *many-worlds interpretation*, and the *de Broglie-Bohm interpretation*, all of which represent attempts to make sense of the quantum world and its peculiarities.

Nessie is looking at me suspiciously. 'Why the interpretations?' My goofy colleagues are in a reflective mood.

It's important to remember that quantum mechanics differs from classical physics. While classical physics builds upon seemingly self-evident truths, quantum mechanics relies on fundamental assumptions, or postulates, that haven't been definitively proven through logic alone. These postulates were formulated to explain the strange behaviours we observe in the quantum world and have been incredibly successful in making predictions. However, the underlying reasons for these postulates remain an active area of research.

Various interpretations are offered by physicists, each vying to explain what the meaning of quantum measurement is.

Collapse of the wave function interpretation

The wave function interpretation, or the Copenhagen interpretation, formulated primarily by Niels Bohr and Werner Heisenberg in the 1920s, is one of the earliest and most widely taught interpretations of quantum mechanics.

At its core, it says that the act of measurement collapses a particle's wave function, determining its specific state, Figure 20. Prior to measurement, particles exist in a superposition of multiple states, and it is only through observation that one particular outcome becomes realised.

One key aspect of the Copenhagen interpretation is the role of the observer and the acknowledgment of inherent uncertainty in the quantum world.

It is reliant on measurement as a fundamental process, leaving the nature of reality unexplained until observed. The idea that the act of observation by a conscious observer creates reality is what causes Dennett to question the nature of truth.

Andrei Linde is a prominent Russian-American theoretical physicist and professor of physics at Stanford University whose work has been fundamental in the field of cosmology.

Andrei suggest a delayed Schrödinger cat experiment.

Iesha takes the radioactive atom onboard a spaceship and flies out to Earth's orbit. Lilly sets up the trigger for the vial of poison on Earth.

And Nessie does what Nessie does best, guards the pack.

In this scenario, if the atom decays, a laser beam is sent by Iesha to the box, the poison is released.

So, now, who killed the cat? Iesha? Or Nessie for looking?

But Nessie does not check the cat for a week.

This experiment has only two possible outcomes. After a week, Nessie either sees a very hungry cat, or a smelly piece of meat. Either the cat was dead a week ago or was alive and got hungry.

Linde says, "Iesha must be blamed here. The cat was dead for a whole week before Nessie looked. But Copenhagen interpretation of quantum mechanics says it is Nessie who collapses the wave function. The interpretation suggest that the cat was either alive or dead all week and there has been no definite state of cat until Nessie looked."

He continues, "Well, this is patently wrong, because from the state of the cat we know the vial was triggered a week ago."

Nessie is blameless. Iesha sent the laser pulse … and she also looks guilty.

Once we establish that Nessie, as an observer, is blameless for the demise of her friend, the cat, then we can answer questions such as, when we observe the Andromeda Galaxy through a telescope, does it imply that the galaxy existed before we observed it?

Copenhagen interpretations says the galaxy was in a state where it could be there or not, but we know everything looks *as if* the galaxy was there before I started observing. *As if* removes all certainty in our knowledge about the universe!

We have never understood what a collapse of the probabilistic wave function actually means. The Copenhagen interpretation may be useful for calculations, but it fails to provide a satisfying philosophical account of the true nature of physical reality.

The fur girls and I think it is just a placeholder for a better theory of particles in physics.

Many-worlds interpretation of quantum mechanics

The many-worlds interpretation, proposed by Hugh Everett III in 1957, provides a radical alternative, where all possible outcomes of a quantum event actually occur, each in a separate branch of the universe. These branches represent different quantum states, and the observer becomes entangled with one specific branch corresponding to the observed outcome.

A multiverse where all possible outcomes occur in parallel realities.

While resolving the probabilistic nature of quantum mechanics, it raises philosophical questions about the nature of reality and the empirical testability of multiple universes.

This interpretation is gaining in popularity, but it makes matters even worse; it is not testable and, therefore, is not a scientific explanation of quantum measurement.

In many-worlds interpretation, there are almost infinite, 10^{122}, universes and in some of these universes Nessie, Iesha and Lilly are playing chase with their aliens.

Pilot-wave interpretation of quantum mechanics

The de Broglie-Bohm interpretation, also known as pilot-wave theory, was developed by Louis de Broglie in the 1920s and later refined by David Bohm in the 1950s.

This interpretation introduces a hidden variable, the particle's *pilot wave*, that guides the motion of particles in addition to the wave function. Unlike the Copenhagen interpretation, de Broglie-Bohm is a deterministic theory that preserves the concept of particles having well-defined trajectories.

In this interpretation, particles are not confined to a probabilistic cloud of possibilities but follow definite paths determined by the interaction of the particle's pilot wave and the wave function.

Here the measurement process is explained as the interaction between the pilot wave and the particle, resulting in the apparent collapse of the wave function.

The de Broglie-Bohm interpretation introduces non-locality, where the behaviour of a particle is influenced instantaneously by changes in its environment, violating the principle of locality in classical physics.

The theory also requires a guiding wave, leading to questions about its physical reality and interpretation.

Measurement problem or observer problem?

Quantum mechanic's focus on measurement means the way we observe affects the very nature of reality and consciousness plays a fundamental role in defining physical reality.

But before jumping to conclusions here, let's remember the measurement problem is specifically about the probability **wave function collapse**, the transition from a quantum system existing in a superposition of possible states to a single definite state upon measurement.

We don't understand it, so we have several interpretations of quantum mechanics which offer competing, unscientific answers to these questions.

And things get worse. If the way we measure, observe and if our consciousness plays a fundamental role in determining the nature of reality, then does the universe even exist independently of our perception?

Or is there absolute objectivity in mathematical and physical models? Without a conscious observer?

All four of us agree with our hero, Roger Penrose, who says this all means there is something wrong with our theory, which might be accurate and precise, but we do not understand what it really means and how it describes the reality we live in.

As Sabine Hossenfelder[3] [4] very eloquently puts it, "the problem in physics is called *the measurement problem* and not *the observer problem* because it is a measurement problem."

Decoherence: does reality need us?

Can the wave function collapse by any other means than by measurement? Yes, it is called decoherence.

The concept of wave function collapse and decoherence in quantum mechanics are both related to the transition from the quantum to the classical world, but they represent distinct processes with crucial differences.

To collapse the probabilities wave function, we do not need an observer at all; quantum decoherence, which refers to the gradual loss of quantum coherence, or loss of quantum state, in a system can happen through interaction with the environment.

Decoherence is a natural consequence of the Schrödinger equation and is fully described by the unitary evolution of the combined system (quantum system + environment). It does not need measurement or a conscious observer.

Even though individual atoms and molecules can exist in superpositions of states, decoherence ensures that entanglement with the environment quickly destroys these superpositions, leaving behind a mixture of classical-like states for the macroscopic observables.

Decoherence explains why macroscopic objects exhibit classical behaviour.

The observer's role in physics: the meaning of measurement

The observer has a role, but it has nothing to do with the wave function. As you see above, we do not necessarily need an observer to move from micro, probabilistic world, to macro, deterministic world.

A conscious observer has nothing to do with creating reality by observing it.

We play a role in selecting the positions of our telescopes, satellites and other apparatuses, which introduces bias into the data we collect. This bias must be taken into account. When we look out into the sky, we can observe certain galaxies and stellar objects, yet there are many that we cannot observe.

If we want to draw conclusions from the statistics we collect, we have to take into consideration factors such as the direction we looked at and the type of data our telescope has gathered. This has nothing to do with consciousness creating the reality of stellar objects and galaxies in the cosmos.

The unitary matrix used for quantum measurement is called the projection matrix. The projection matrix for a measurement of a qubit is a 2x2 matrix that has the following form:

$$P = |0\rangle\langle 0| + |1\rangle\langle 1|$$

What does this mean? This is the mathematical definition of quantum measurement and decoherence. It is any activity that leads to the collapse of the wave function in quantum mechanics.

Turns out, even a random particle bump can cause a quantum system to change its state; the wave function collapses without a conscious observer needed.

I explain to my fur friends, 'It's just how the messy world interacts with delicate quantum stuff. A cosmic ray hitting some sand is measurement. It doesn't need a scientist with a clipboard.'

Some physicists get so jumbled in the maths, they think *we* make electrons real by looking at them.

The universe is well capable of changing the state of a quantum system without us being at the centre of universe.

It needs no consciousness. It needs no observer.

Consciousness observer as the creator of physical reality?

Eugene Wigner,[5] a pioneer of quantum theory, argued that conscious perception is integral to the measurement process. This view, aligned with the Von Neumann–Wigner interpretation, has garnered support from those concerned about the paradoxes inherent in a non-conscious universe randomly selecting one reality out of many possible ones. The theory suggests that consciousness is not merely an emergent property but a fundamental aspect of the mechanism by which the universe operates.

However, the idea that consciousness is necessary to shape physical reality presents a host of issues. Defining consciousness rigorously is challenging, making it difficult to test whether a conscious observer is truly required for quantum measurement.

Furthermore, the very notion that consciousness has a privileged influence on the behaviour of matter seems antithetical to what we understand of the laws of physics.

Quantum effects have been a fundamental aspect of the universe for 13.8 billion years, long before our arrival 250,000 years ago.

At the Big Bang, there was no consciousness. Just pure energy.

Quantum measurement does not require human-made apparatuses to occur; it is merely a function of the decoherence of quantum states.

Let's remind ourselves, measurement is just maths. It is not, in itself, reality.

Physics is an approximation

Physics seems to be, at best incomplete, and at worst limited in its ability to fully describe reality.

The world that we experience, the deterministic macro world, seems nothing like the probabilistic micro world that it is made up from. Which is real? Why can we not relate to a probabilistic world?

Roger Penrose says: "There is something wrong with our physics." Nima Arkani-Hamed agrees: "There is something wrong with our physics of general relativity and quantum mechanics."

The race for a unified theory that seamlessly merges the macroscopic world of general relativity with the microscopic domain of quantum mechanics is on.

To be clear, we have a model that describes atoms, molecules, photons and electrons, i.e. the constitutes of our world, very accurately. But these laws are strikingly different from those that dictate the motion of planets, galaxies and even us humans and dogs.

Our theories, while undeniably powerful, are riddled with inconsistencies. They offer contrasting, even contradictory, descriptions of the fundamental nature of reality. Their description of nature of time, nature of physical reality, determinism and probabilistic world are not in line with our experiences.

The conscious machine and physical reality

A conscious machine, capable of experiencing and interacting with the world in a manner analogous to a human, would be a watershed moment. If the *consciousness causes collapse* interpretation holds true, it would introduce a degree of control over the behaviour of quantum phenomena that has eluded physicists thus far.

We could potentially design experiments where the conscious machine acts as the sole observer. Its internal state would hold profound ramifications for the outcomes of quantum measurements. We might discover whether subtle changes in the machine's conscious experience could directly influence the collapse of the wave function. Such experiments could unveil surprising links between consciousness and the very fabric of reality.

The potential impact of a conscious machine would reach far beyond pure theoretical physics. In quantum computing, the superposition of states is a crucial resource. Building a quantum computer that interacts with and is *observed* by a conscious machine could allow for new classes of algorithms that incorporate the machine's conscious state directly into computations. Such a hybrid could enable unforeseen breakthroughs.

Naturally, the development of a conscious machine leads to a host of ethical concerns. We would need to determine its rights and whether it could be considered analogous to a person. The very act of manipulating a conscious entity for scientific or technological purposes would raise complex moral questions about the nature of sentience and agency.

Nessie, Iesha and Lilly think, 'That is of course if the collapse of the wave function is due to measurement by a conscious observer!' Oh, their walnuts are on fire today.

The conscious alien and physical reality

If the laws of physics are universal, and consciousness arises naturally in complex systems, then *consciousness causes collapse* interpretation may face additional challenges.

The existence of intelligent alien life elsewhere in the cosmos is a tantalising possibility that carries profound implications for quantum theory as well. If alien civilisations developed sufficiently advanced technology to understand quantum mechanics, and if such aliens possess consciousness comparable to our own, their role as conscious beings within the universe takes on even greater significance.

The existence of consciousness, having evolved independently in other locations within the universe, would offer support for the notion that consciousness is a natural outcome of complex physical systems.

This could bolster our arguments that a consciousness-dependent interpretation of quantum mechanics is not necessary for the creation of physical reality.

As far as we know, everything we detect and everything we see in the universe is similar to our physical reality on Earth. If conscious aliens created the physical reality in a corner of the universe, what is the probability that it would look exactly as is in our locale?

The implications of both conscious machines and conscious aliens remain firmly in the realm of speculation, contingent upon both technological and scientific breakthroughs that may be far in the future, however, we think they form a great thought experiment against the collapse.

In a nutshell

The whole universe is not waiting for us to observe quantum states collapsing for it to become real.

Reality is better defined by a broader set of principles, with our conscious perception tailored to navigate through it.

Consciousness is not a property of reality and time in physics. Reality is not a property of consciousness. We became conscious to better cope with physical reality.

Our physical reality is not well captured within the confines of mathematics and physics models. Physical reality is most reliably captured by our perception.

There's a profound gap in our theories of physics.

Chapter 10:
Are We In A Computer Simulation?

Sci-Fi vs reality

Imagine a world where everything you see, feel and experience is actually a computer program. Could this ostensibly far-fetched idea be true?

The Evolutionary Lens suggests consciousness arose to better navigate the physical world, but what if our reality isn't what it seems?

This chapter explores the simulation hypothesis challenging the very basis of reality. Could we be digital characters unknowingly trapped within a hyper-realistic simulation, and is there any way to tell?

Information is physical

The notion that we might exist within a computer simulation forces us to confront unsettling questions about physical reality. Could the limitations we perceive in our world, speed limits, finiteness of resources, even the laws of physics themselves be artifacts of a programmed environment?

I explain to my collaborators that the question of whether we live in a computer simulation explores an important concept regarding the nature of information.

David Deutsch[1] says that "there's a notorious problem with defining information within physics, namely that information is purely abstract and the original theory of computation, as developed by Alan Turing, regards computers and the information they manipulate purely abstractly as mathematical objects."

When Babbage and Turing first uncovered the theory of computation in the 1930s, its connection to physics was not immediately understood. Initially, it was conceived as a branch of mathematics aimed at exploring mathematical proofs.

The foundation of information theory was established on the assumption that a specific abstract entity, the Turing machine, had the capacity to represent all possible computations.

We've discovered that information theory is deeply linked with physics, evolving into the branch of quantum computation. We can use particles and forces from physics to represent and process complex information. It's a deep and basic part of how we use nature to do calculations.

With the advent of quantum computers, it becomes apparent that a physically real universal computer surpasses the capabilities of a classical Turing computer.

Quantum computers take advantage of our deepest understanding of the physical world, quantum theory, therefore allowing these computers to solve complex problems that are either theoretically or practically impossible to calculate in traditional computers.

For example, Nessie, Iesha, Lilly and I, started this morning with some kind of electrochemical signals in our brains, which were then converted to signals that travelled out, generating sound waves which were then turned into mechanical vibrations of ear drums, back into electricity and so on.

During this morning's walk and talk, the concept of information manifested in a variety of physical forms that adhere to different physical principles, from electric signals, to wave propagation, to harmonic isolation and electricity.

Deutsch[2] explains, "To convey the essence of this process, we must acknowledge the one element that has consistently persisted throughout, the information itself. This is unlike energy, momentum, or gravity, which can change in specific circumstances."

He continues, "Mathematicians still do not recognise that information itself has physicality; abstract computers do not exist. Computation is a process that necessitates a physical form."

Physicists have always considered information to be a physical quantity.

Information itself is abstract. It can be represented or stored in various physical forms (a book, a computer file, brainwaves) without its essential nature changing.

A computer that could build a whole universe inside itself, would still have to play by the same rules that govern our universe. Think of these like the unbreakable rules of a video game – even the programmer has to follow them.

That means this simulator-computer would also have to exist

within the universe it's trying to recreate, just like everything else.

That means aliens so advanced they could create our entire universe on their computer have mastered the deepest secrets of how the universe works; down to the very last detail. But there's a catch ... in Chapter 6, we learned about a concept called *Darwinian Time*. This seems to put limits on how quickly any civilisation, even an incredibly advanced one, could progress.

Iesha thinks, 'Well, the computer and the alien can be outside of our universe, simulating our universe.'

I explain, 'Oh, wait a minute, other universes, if they exist, would not necessarily have our laws of physics.' I go on to explain why this would not make sense.

You do not have to worry about the complexity of the three paragraphs below. I just am trying to show doggies how bizarre this proposition is.

Let's imagine Zaza is a very clever alien outside of our universe who built an amazing machine to create her own personal playground universe in some software. By the way, Zaza was our first adopted dog and Nessie's best friend. She played Zeus in our house.

Zaza decides to program a computer simulation with 10^{500} stable, lowest-energy state vacuum. She then chooses one of these stable vacuums, improbably the right one, to create a model of the universe to simulate.

Now if it were us, our simulated universe would be relatively simple. Our electron mass would be equal to the proton mass, equal to the mass of the W Boson, equal to the mass of Z Boson; all masses would be the same. All coupling constants would be the same. This universe is simple to program. Just a few bits of information and we would have our simulated universe.

But oh, no, Zaza is clever, she creates her universe in a very weird state. In her sim universe, an electron is 1,836 times lighter than a proton, which is 85,987 times lighter than W Boson, which is 1.13 times more massive than Z Boson, all coupling constants are different …! Sounds familiar? This is the description of our universe!

Zaza's universe is incredibly complex with a bizarre relationship between the fundamental particles.

Everything is so strange in Zaza's simulated universe.

Her simulated world is so intricate that its underlying code would have to be far more complex than the physical laws that shape our seemingly boundless universe.

I continue, 'Essentially, claiming that the universe runs on a computer, based outside of the universe, is akin to asserting that Zeus created a computer to simulate the universe as a program! Wouldn't it be simpler for Zeus to just create the universe?

'Zaza is designing a universe sim more complex than the real thing. Why bother when a simple system would be easier to create or maintain? All she would have to do is set the initial conditions and let it play out.'

Despite our extensive exploration of the universe, we have yet to discover any definitive evidence suggesting the presence of a simulation or modelling device.

Furthermore, if this hypothetical computer was located outside our universe, it would remain entirely beyond our detection capabilities. Without the ability to devise experiments capable of probing beyond our universe, any attempt to verify its existence would be futile, rendering the endeavour essentially meaningless.

A concordance check: is our universe just a giant video game? And who's holding the controller?

While the simulation idea is tantalising, there's simply no way to scientifically test if it's true. The concordance model highlights that while our understanding of the universe is remarkable, some questions may be out of reach for now.

Physical laws as code: The concordance model emphasises the consistency of the universe's fundamental laws. If we are in a simulation, these laws must be seen as the programmed *rules* of our simulated reality. The model prompts us to question: could those *rules* be altered or reprogrammed by the hypothetical simulator?

Universal limits: The concordance model acknowledges limitations in our understanding of the cosmos. Perhaps the reason we haven't detected the simulation program running our universe is that there is a limit to what we can detect in our visible universe.

In a nutshell

Are we in a sim? Or did the universe just pop into existence?

Honestly, both ideas raise mind-boggling questions.

The sim idea feels overly complicated; way too much work for whoever made us, aliens OR Zeus. While it's a fun puzzle, science can't really answer it. But exploring the idea does force us to think deeper about what *reality* even means.

Chapter 11:
The Nature of Time In Physics

Theory of special relativity suggests that the distinction between past, present and future is relative to different observers in our universe. This raises the possibility of a model of our universe, called *block universe*, where all moments in time exist simultaneously, though our experience remains bound to a linear progression.

If that's true, how does anything ever change? And what does it mean for us?

Our experience of time seems at odds with the rigid *timelines* of physics. How do we reconcile these?

Philosophers like Henri Bergson argued that for us time is fundamentally subjective but then Einstein flipped our understanding of time's objective nature.

From the physics of time to the neuroscience of perception, this chapter explores the relationship between time and our conscious minds.

Surround yourself with fur friends; you won't experience a dull moment!

The concept of time is a mashup of physics, philosophy and neuroscience.

During our walks, I frequently remind Nessie that we are losing daylight. She insists on sniffing every blade of grass and greeting every living creature we encounter in the park. It is both cute and annoying to the same measure.

It's amazing how differently we experience the passage of time. Somehow, time goes slow for me and fast for her. She is having fun.

Experiential aspects of time as perceived by organisms in their slow-moving and, therefore, non-relativistic experience is best explored in the context of IIT; our subjective experience of the passage of time is shaped by how our brains handle information, memories and sensory inputs.

Lilly is confused. 'Why do you say that time, as we perceive it in our minds, seems to differ from time in the physical world? What's the difference?'

I suggest an experiment to her: 'Think about the *now* moment in your head. It's strange, isn't it? It's elusive; you can't pin it down. Is it now? Or now? Or now?'

The sense of a slippery *now* is key! That's where physics and our subjective experience of time clash. In physics *now* is just a point on the chart. Absolutely clear what now is. You can point to it and say, 'There, *now* is just there.'

Relativistic nature of time as presented by Einstein is still not well understood. To be clear, general relativity has been tested for 100 years and for all practical purposes seems to be accurate. We rely on it for the accuracy of all our electronic gadgets. What we mean by *not well understood* is that the fundamental nature of time is not clear to us.

Stephen Wolfram[1] says space is a property of time and time is fundamental. In essence, time creates space. Our universe is fundamentally single dimensional and space is created through passage of time.

Wolfram has demonstrated this phenomenon computationally.

But general theory of relativity tells us time is a fundamental property of the universe and one of the dimensions of spacetime we live in.

Which is it? Who is right? Is time a fundamental property of spacetime, or is space an emergent property of time?

The early 20th century saw a peak in the philosophical debate over the nature of time, marked by the notable disagreement between Albert Einstein and Henri Bergson.

Henri-Louis Bergson (1859–1941),[2] a French philosopher of *time*, argued for an intuitive and experiential understanding of time. He said that the scientific representation of time neglected the richness of human consciousness and the subjective flow of duration.

Bergson's view on the nature of time is quite unique and stands in contrast to the traditional scientific and philosophical conceptions. He offers a critique of the *spatialisation* of time, saying that it is fundamentally different from space and cannot be understood through spatial metaphors.

"Our lived experience of time, characterised by continuous flow, interpenetration of past, present and future, and constant change. It's qualitative, non-quantitative, and cannot be broken down into discrete units."

Bergson argues that attempting to understand time through spatial metaphors like points and lines creates a fundamentally flawed image. Time is not a line of moments; it is the continuous unfolding of experience itself.

Clock Time (Durée homogène) is the time of science and mathematics; a linear, homogeneous, divisible and reversable abstraction based on spatial models. It is a useful tool for practical purposes but ultimately fails to capture the true nature of time as we experience it.

The spatial view of time leads to a false conception of the past as separate and fixed, the present as fleeting, and the future as entirely open.

In real physical duration, past, present and future are continuous.

Bergson argued for a more intuitive and experiential understanding of time. He contended that the scientific representation of time neglected the richness of human consciousness and the subjective flow of duration.

The clash between these perspectives underscores the tension between the physical and mental aspects of time, highlighting the challenges in reconciling these seemingly divergent viewpoints.

The more recent version of this is Tononi's proposal that time plays a crucial role in information integration. According to IIT, consciousness relies on the ability of different brain regions to exchange information and become co-dependent on each other's activity.

This requires a temporal, or time-related, order, where information flows sequentially from one region to another, allowing for the build-up of integrated information over time.

He goes further and suggests that time is not an objective property, existing independently of brain. Instead, he proposes that our experience of time arises from brain's own internal information-processing dynamics. The flow of information within brain, with its sequential and causal relationships, creates the subjective sense of time passing.

Nessie, Iesha and Lilly think, 'The implication that time arises from brain's internal information processing is stretching things too far.' I agree.

The proposition that by imagining time we create time cannot be valid. Time existed in the universe before we appeared to take the accolade for inventing time in our heads.

A unanimous show of paws and hand reveals that we all agree Bergson has made a valid point.

In the physical world, time has a direction from the past, to the now and to the future. Equations of motion in physics and quantum mechanics can be wound backwards.

Although equations of motion can be rewound to trace the path, it does not mean the Earth can change direction and trace its path backwards.

These equations convey that the physical world operates on deterministic principles, allowing us to retrospectively trace events and forecast their future trajectory. Similarly, the concept of time's arrow in physics suggests that time moves in a singular, forward direction.

The clash between Bergson's and Tononi's perspectives and that of Einstein's underscores the tension between the physical, biological and mental aspects of time, highlighting the challenges in reconciling these seemingly divergent viewpoints.

Time dilation: go faster to age slower

In the theory of general relativity, space and time are intertwined and curved by gravity. This theory also predicts various phenomena related to time, such as gravitational time dilation (time runs slower in stronger gravitational fields) and gravitational waves (ripples in spacetime caused by massive objects accelerating).

These predictions have been confirmed through observations of various astronomical phenomena such as binary pulsars and the recent detection of gravitational waves.

Imagine my two friends, Iesha and Lilly, living on Earth. Iesha hops on one of Elon Musk's new generation of super-fast spaceship and zooms around the universe for a while, then returns home. When the two reunite, something astounding happens: Iesha is younger than Lilly! How is this possible? The answer lies in a fact about our universe revealed by Einstein.

In special relativity, time dilation refers to the phenomenon where time runs slower for an object moving at high speeds relative to another observer. This means that, for the moving object, time intervals between events will be longer compared to the same intervals for the stationary observer.

Nessie is wondering, 'The squirrel is faster than me, so does she get younger and younger relative to me as I chase her?'

'I am afraid so Nessie', but I assure her by a very tiny amount at their running speeds. Nessie is standing still and staring at the squirrel who is running at 30 kph. The squirrel's speed is just 0.0000278% of the speed of light. Plugging this value into the time dilation equation, we get a time dilation factor of about 10^{-16} sec.

The difference in experienced time is extremely small. This is trillions of times smaller than a second and beyond the measurement capabilities of most current technology.

Time dilation isn't just about clocks changing, it's like ... two friends ageing at different speeds!

I continue, 'For all practical purposes, Nessie, the squirrel's age remains the same as yours, even with its impressive running speed.' The time dilation effect at everyday speeds is simply too minuscule to be perceived.

This concept isn't just theoretical. Extremely precise clocks have confirmed this effect.

How to be seen to live your life backwards: relativity of simultaneity

Remember way back in the introduction, we talked about George Carlin and how he can be granted his wish to live backwards in Einstein's universe? That was how this whole book started.

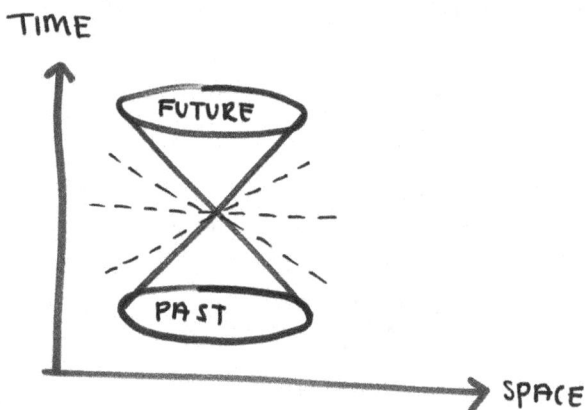

Figure 22: Light cone

Here and Now is the centre of Figure 22. Everything in the bottom triangle is definitely in the past, and everything in the top triangle is definitely in the future.

But things outside these two triangles can be either in the past or future or present, depending on how fast you're moving. The dashed lines are examples of different *'nows'*. In this diagram, time points up and space points left/right.

According to special relativity, if we were to move at a speed that is comparable with that of light, then events that seem to us to be simultaneous would generally not seem to be simultaneous to some other observer with a different velocity.

Moreover, the velocities would not even have to be very large if we are concerned with very distant events.

In the universe, there's no single *now* for everyone. Two doggos moving at different speeds would disagree about what events across the cosmos are happening at the same moment. Their view of time is literally different.

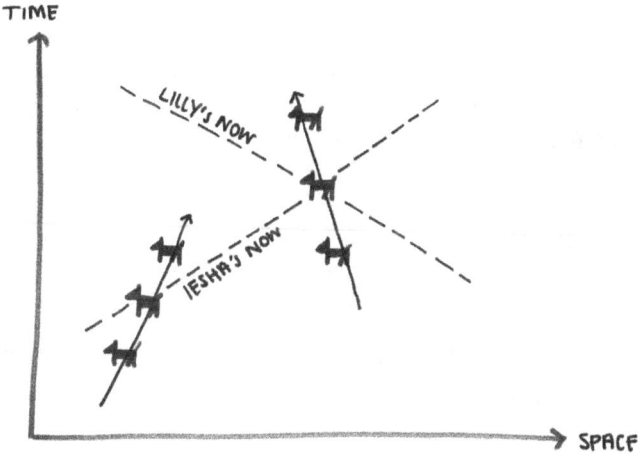

Figure 23: Iesha's future is Lilly's now

By moving fast, Figure 23, and in different directions, Lilly and Iesha have different *nows*. In this diagram, Iesha's *now* includes Lilly at some particular moment, but for Lilly that moment happens at the same time, same *now*, as a time in the *future* for Iesha.

Carlin has a chance to live his dream, to be seen to live backward in space, for observers at the right speed and at the right place.

Entropy and arrow of time: macroscopic vs microscopic view

Where does the physical arrow of time come from?

In classical physics, time reversibility is inherent in the fundamental equations that govern the behaviour of physical systems. Newton's laws of motion and other classical equations of physics are time-symmetric, meaning that they are equally valid whether time progresses forward or backward.

This symmetry is a consequence of the deterministic nature of classical mechanics, where the future behaviour of a system is uniquely determined by its initial conditions.

In classical physics, if you know the positions and velocities of all objects in a system at a particular moment, you can, in principle, use the laws of physics to predict their positions and velocities at any past or future moment.

We can trace back where planets have been, where they are now and where they will be in the future.

This reversibility stems from the fact that classical physics is time-reversal invariant.

This means that if you were to reverse the direction of time in a hypothetical scenario, where all the positions and velocities of objects are reversed, the equations of motion would still be valid, and the system's behaviour would be consistent.

Let's look at this in more depth: one piece of evidence supporting the Big Bang theory is the redshift of galaxies. This concept, known as Hubble's Law, was introduced by the astronomer Edwin Hubble in the early 20th century and has since become a cornerstone in our understanding of the expanding universe.

If the universe is currently expanding, hence the red shift, then we know from the laws of classical physics, in the past it must have been much smaller and denser. The idea that the universe originated from an extremely hot and dense state, commonly referred to as the initial singularity, is a direct result of time invariance in physics.

Time-reversal invariance has given us the ability to estimate the age of the universe at 13.78 billion years old, with an uncertainty of only about 20 million years.

This estimate comes from data collected by the European Space Agency's Planck mission, which observed the Cosmic Microwave Background (CMB) radiation, a faint afterglow from the Big Bang.

The resolution of the arrow of time in the context of quantum mechanics, however, is still ill-defined.

Quantum mechanics also possesses time-symmetric equations. The Schrödinger equation, the foundational equation of quantum mechanics, is also time-reversible. The evolution of a quantum state, described by the wave function, is deterministic and can, in principle, be run backward in time.

However, the issue of time irreversibility arises when considering the process of measurement in quantum mechanics. The act of measurement collapses the wave function into a specific property of that system and has a definite, unchanging value, i.e. *eigenstate,* and this process is considered irreversible.

Once a measurement is made, information about the probability wave function of the system is lost, leading to a definite outcome. We introduced this as the measurement problem in quantum mechanics.

While the fundamental equations are time-symmetric, the irreversibility in the outcomes of measurements has led to the emergence of the arrow of time at the macroscopic level.

Entropy and the arrow of time

Both classical physics and quantum mechanics have time-symmetric equations at their core. The challenge lies in understanding the emergence of arrow of time.

Some physicists argue that there is a profound relationship between *entropy* and the arrow of time.

In the field of thermodynamics, entropy defines the degree of disorder within a system. As systems evolve, they tend to move towards states of higher entropy, mirroring an inexorable trend towards chaos and randomness.

The second law of thermodynamics encapsulates this tendency, asserting that in any isolated system, entropy either remains constant or increases, but it never decreases.

The arrow of time represents the unidirectional flow of events from the past through the present and into the future. Unlike the symmetrical laws of classical physics, the arrow of time introduces an inherent directionality to the temporal progression of the universe.

This arrow aligns itself with the increase in entropy, defining a universe where systems move from ordered states to states of greater disorder.

To grasp the concept of entropy, it's essential to distinguish between the macroscopic and microscopic levels. At the macroscopic level, entropy is often associated with observable changes, like the melting of ice or eggs falling off a table and breaking.

On the microscopic level, it relates to the statistical distribution of particles and their energy states. In the case of a gas, Figure 24, particles exhibit random movement with varying speeds and directions. When the gas expands, the particles spread out into a larger volume, introducing increased uncertainty in their positions. This heightened randomness corresponds to a rise in entropy.

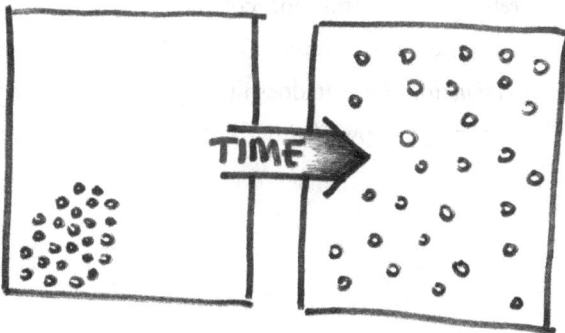

Figure 24: Entropy and arrow of time

In Figure 24, the left side shows numerous gas molecules confined to one corner of a box, a state of low disorder or low entropy. Over time, if the box remains undisturbed, the gas molecules disperse throughout the box due to their random motion, leading to a state of increased disorder or high entropy.

This transition illustrates the second law of thermodynamics, which states that systems tend to evolve from configurations of lower entropy to those of higher entropy, as depicted by the change from the left to the right side of the figure.

The reverse process, moving from the high-entropy state back to the low-entropy state, is exceedingly rare or practically impossible.

This is the arrow of time – we think!

Is entropy really the answer?

I really tried hard to persuade Nessie, Iesha and Lilly that entropy defines the arrow of time, and our minds create internal and personal models of time appropriate to our locale.

But I can tell they are still not convinced. They think, 'This is correlation and not causation: while entropy offers a potential *direction* for time, it doesn't fully explain the nature of time itself, or even the arrow of time.' Clever girls!

Entropy can also decrease!

So far, we have seen in physics the consensus is that the second law of thermodynamics is the definition of arrow of time.

We are just not convinced. There is no evidence to suggest arrow of time is caused by increase in entropy. This explanation is deeply unsatisfactory.

My collaborators think, 'What is the proof that universal entropy continues to increase when in specific locale decreases? How would you even measure universal entropy after each event?'

I say, 'Ohhh, that is so clever!'

'Imagine your messy pen, with all your toys out. Laws of physics say entropy (messiness) tends to increase over time … yet, with effort, you can clean it. Entropy isn't a force of nature, it's a pattern.'

The stunning Hubble Space Telescope's images of stellar nurseries reveal a fascinating process, the formation of thousands of new stars. Formation of stars represent increased order, i.e. decrease in entropy.

Plants convert sunlight energy into complex molecules such as carbohydrates, decreasing the entropy of the surrounding system. All living things maintain order and complexity by consuming energy and expelling waste, creating local order.

Computers and brains store and process information, creating localised order from seemingly random input.

In these and many other such examples, we can see that the argument *that the total entropy of the universe still tends to increase* is a highly unsatisfactory definition of arrow of time.

Iesha thinks, 'This whole entropy always increasing thing; that's the whole universe, right? How do we even prove something like that?' I tell her, 'That is the problem; we cannot.'

The interplay between entropy, the second law of thermodynamics, and the nature of time is nebulous, imprecise, vague and highly unsatisfactory.

Correlation, not causation: Entropy doesn't fully explain the nature of time itself.

Difficulty of measurement: Although there are countless experiments that consistently demonstrate entropy increasing in closed systems, directly measuring the total entropy of the *entire* universe is an impossible task.

We simply can't observe and catalogue every single particle and interaction. The position that the entropy of the system, i.e. the whole universe increases, is not measurable or scientific. It is a hunch!

While it's tempting to connect entropy with the flow of time, the relationship is nuanced and needs deeper investigation.

Physics provides profound insights into the workings of the universe, yet the origin and fundamental nature of time remain elusive.

If Stephen Wolfram is right and time is the engine that creates space, then perhaps our current understanding of the physical world is incomplete. Wolfram's proposition that time is fundamental and can be proven computationally may be our most promising definition of time to date.

In a nutshell

While physics provides a detailed and predictable understanding of time's flow, it grapples with explaining its fundamental origin. Furthermore, from the vantage point of an observer traveling at near-light speeds, the concepts of past, present and future become fluid and relative. It appears that these temporal categories coexist as blended and laced layers within the fabric of spacetime itself.

Chapter 12:
The Brain's Clock

The Evolutionary Lens: universal clock synchronisation

In Chapter 1, we demonstrated that a single external physical reality serves as the training data for the neural networks of countless organisms on Earth. Time, a physical universal constant, is a fundamental aspect of this reality. Together with space, it forms the fabric of the universe, governing the tempo of the cosmos and the cycles of our planet.

As life evolved on Earth, it became intrinsically linked to the planet's rotation, characterised by the rise and fall of the sun, serving as a primary timekeeper. This physical clock shaped the biological rhythms of all organisms. *Circadian rhythms* are a 24-hour internal clock which regulates many biological processes in organisms, such as sleep-wake cycles and hormone fluctuations. At the core of these rhythms lies the *suprachiasmatic nucleus* (SCN), a tiny brain region in the hypothalamus. The SCN acts like the body's main clock, using light signals from the eyes to match our internal rhythms with the day and night cycle.

While Earth's rotation provides a physical beat, each organism perceives its own clock. Evidence suggests that time perception is not uniform across species; it's a tailored experience, honed by evolution to meet the specific demands of each ecological niche. For instance, a humming bird's perception of a fleeting moment differs vastly from a tortoise's leisurely pace. This diversity reflects the adaptability of time perception, shaped by natural selection to enhance survival and reproduction. It underscores the complex interplay between evolution, biology, and the subjective experience of reality.

Therefore, our perception of time is anything but constant. It warps and bends, influenced by our biology, our experiences, and our brains.

Time To Reach Squirrel (TTRS) And Time To Doggo Arrival (TTDA)

At this morning's walk the topic of conversation was Time to Reach and catch the pesky Squirrel (TTRS) and Time for fast Doggo to Arrive (TTDA). These concepts are crucial for understanding predator-prey dynamics, as both predator and prey must assess physical reality, including time and space to make split-second decisions.

The squirrel's perception of time is not an objective measurement but a subjective construct augmented by memory and attention, often deviating from the physical passage of time.

This raises an important question about the utility of subjective time perception in survival.

Consider a squirrel evading a dog. The squirrel doesn't need to calculate the TTDA to the millisecond; it simply needs to sense danger and react faster than the dog. Her internal clock, while subjective, is an adaptive mechanism that prioritises immediate survival.

Evolution doesn't select for perfect timekeepers; it favours time perception mechanisms that enhance survival and reproduction. Therefore, while an organism's internal clock might not be perfectly accurate, it plays a crucial role in the survival toolkit, providing a critical advantage in life-or-death situations.

"Ticking away the moments that make up a dull day" - Pink Floyd, 1973

Neuroscience reveals a complex network of brain structures that govern our perception of time. Emotions, processed by the *amygdala*, interact with the prefrontal cortex and other temporal processing regions, shaping how we perceive time during emotionally charged events.

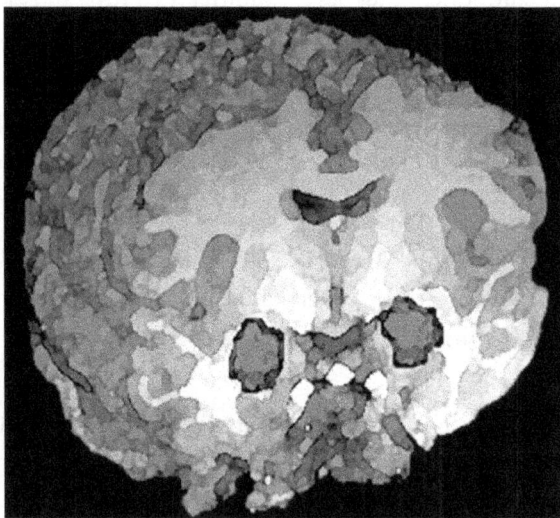

MRI view of the amygdala, a pair of almond-shaped structures within the brain's temporal lobes

Beyond the influence of emotions, specific neural mechanisms play a crucial role in timekeeping. *Neural oscillations*, rhythmic patterns of brain activity generated by synchronised firing of neurons create a kind of internal metronome. Different frequencies of these oscillations contribute to different aspects of time perception.

Theta oscillations (4-8 Hz), associated with relaxation and daydreaming, are involved in estimating the duration of events.

Gamma oscillations (30-80 Hz), linked to heightened attention and focus, help us perceive the precise order of events.

Interestingly, the speed of neural oscillations can be influenced by body temperature, metabolic rate and neurotransmitters like dopamine. This could explain why time seems to slow down during a fever or accelerate with heightened arousal.

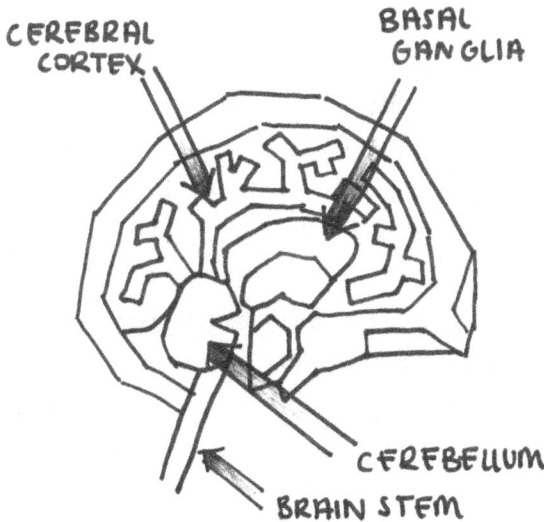

Figure 25: Main brain regions involved in the neural network of time perception

Emotions also play a role in time perception; a squirrel facing a dog might experience a surge of adrenaline, making seconds feel longer as its brain prioritises threat detection. Conversely, the joy of discovering food can trigger dopamine release, potentially leading to a feeling of time compression.

The brain has the ability to adjust its internal clock based on context and attention level. This adaptability is crucial for survival, allowing organisms to accurately perceive and respond to the ever-changing world around them.

The illusion of *now*

Building on our understanding of time perception, neuroscientist Karl Friston's concept of active inference[1] offers further insights. He suggests that the brain operates as a prediction machine, constantly generating predictions about the future based on sensory input and past experiences.

When sensory input aligns with our predictions, we experience a sense of continuity and flow. However, when there's a mismatch, the brain updates its model, creating a perceived shift in the present moment.

Similar to a movie director, the brain constantly predicts the next scene based on the script and past footage. When the actual scene matches the prediction, it feels smooth and continuous – that's our "now".

Our "now" is not an objective reality but a subjective experience, a dynamic window through which we perceive the world. It integrates sensory information, memories, and predictions into a coherent narrative that guides our actions. As the brain navigates the possibilities of the future, the interplay between prediction and perception contributes to the sense of real time.

Different neural systems process information at varying speeds, yet the brain seamlessly integrates these inputs, maintaining the illusion of simultaneity. This ability highlights the brain's capacity to reconcile temporal disparities and construct a cohesive real-time narrative.

From an evolutionary perspective, the present moment is less about precise timing and more about survival. Organisms don't need to contemplate the abstract nature of time; they need to react to their environment in real time. The present moment is a window of opportunity to make split-second decisions – to flee, to feed, to mate. The brain's ability to create a sense of "now," even if it's slightly delayed and constantly shifting, is a fascinating adaptation for survival.

Furthermore, our ability to recall the past and anticipate the future hinges on the intricate connection between memory and temporal processing. This temporal anchoring contributes to the sense of real time, allowing us to situate ourselves within the continuum of past, present, and future.

The past is not a static repository but a dynamic reconstruction influenced by our present perceptions. Similarly, the future is shaped by our current choices and anticipations. Our brains actively construct our experience of time, weaving events into a narrative that shapes our understanding of ourselves and the world around us.

In a nutshell

Our bored brains, our excited brains — all experience time differently.

Time, as perceived by our brains, is a delicate balance of biological rhythms, neural oscillations, emotional states and memory processes. It is a subjective reality, a personal narrative shaped and personalised by each organism, but ultimately defined by the constraints of physical time.

Chapter 13:
Epilogue

The story of Nessie, Iesha, Lilly and our early morning walks, and all our chats along the way, is absolutely true. They earned their place in this book. I discovered that often my ideas sounded so much better in my head before I ran it by them.

This is how my fur companions became my unwitting collaborators as I grappled with complex ideas. It turns out the best way to explore reality might be to ditch textbooks and go for a walk.

On one such a walk we hit upon an idea: what if we could conceive an instrument to interrogate physical reality itself?

Forget about theories. Let's get experimental.

Over many walks and years, we realised the best instrument already exists: our brains, shaped and honed by the evolutionary process.

A wealth of experimental data reveals an intriguing pattern: most creatures with nervous systems seem to navigate their surroundings with a surprising level of *understanding*. The ability to interpret their environment, so vital for survival and reproduction, hints at the possibility that some degree of consciousness might be a common thread, woven into the very fabric of life itself.

Evolution forces all creatures, trillions upon trillions of organisms, to solve the same basic problems. Converging on similar solutions hints at a single, shared reality guiding life. What better tool to use to probe reality than one which is honed for over 5 billion years, tested to destruction, literally, by physical reality itself and is present in countless roaming animals on planet Earth?

Complex behaviours, elaborate social structures and problem-solving abilities observed in diverse creatures hint at a deeper inner experience than we often acknowledge.

With The Evolutionary Lens as our apparatus, we've peered into the depths of physics, reality and time. What emerges is rather an appreciation for all the questions that, hitherto, remain unanswered; impatiently waiting for philosophy and science to catch up.

Consciousness, despite its seemingly accidental origins, may be the universe's most elegant way of understanding itself.

What next?

Consciousness, far from being a supernatural anomaly, appears deeply rooted in the physical universe. This is not to reduce the richness of our inner lives to mere calculations, but to recognise the beauty in how simplicity begets complexity.

Our scientific discoveries, our philosophical musings and even our works of art are manifestations of consciousness exploring its own depths. There is an undeniable wonder in this recursive loop, this ability of the universe to turn back and reflect upon itself.

Our brain is both a product of the universe and the lens through which we capture it.

This perspective promotes a profound feeling of connection to the universe, and a deep responsibility.

Evolution has given us reason, compassion, and the capacity to make ethical choices. It is our duty to build a world that values all life, and nurtures consciousness in all its forms, both biological and artificial.

Armed with slices of cheese and pears, this is the topic we are currently exploring on our walks.

References

Introduction

1. George Carlin – I want to live my next life backwards, Class Clown (1972) and *Occupation: Foole* (1973).
2. Susan Blackmore. *Consciousness: An Introduction* (2nd ed., 2010).
3. Peter Woit. Not Even Wrong: *The Failure of String Theory and the Continuing Challenge to Unify the Laws of Physics*.
4. Richard Dawkins (1976). *The Selfish Gene*.
5. Tononi. Phi: *A Voyage from the Brain to the Soul* (2012).
6. Koch. *Consciousness: Confessions of a Romantic Reductionist* (2012).

Further reading

Blackmore, Susan. *The Meme Machine* (1999).

Blackmore, Susan. *Consciousness: A Very Short Introduction* (2004).

Blackmore, Susan. *Ten Zen Questions* (2012).

Blackmore, Susan. *Beyond the Body: An Investigation of Out-of-the-Body Experiences* (1982).

Blackmore, Susan. *The Adventures of a Parapsychologist* (1986).

Blackmore, Susan. *Consciousness: An Introduction* (2nd ed., 2010).

Hume, David. *An Enquiry Concerning Human Understanding.* Oxford: Clarendon Press 1975.

Koch. *Consciousness: Confessions of a Romantic Reductionist* (2012).

Norton, David Fate. *David Hume: A Life.* London: Yale University Press (2000).

Peirce, Charles Sanders. The Fixation of Belief. T*he Essential Peirce,* Vol. 1. Bloomington: Indiana University Press (1998).

Popper published his thoughts on falsifiability in his book *The Logic of Scientific Discovery,* which was first published in German in 1934 and then in English in 1959. Stanford Encyclopedia of Philosophy: https://plato.stanford.edu/entries/popper

Quine, W. V. O. The Problem of Induction. *The Ways of Paradox* and Other Essays. Cambridge, MA: Harvard University Press (1976).

Salmon, Wesley C. *The Foundations of Scientific Inference.* Pittsburgh: University of Pittsburgh Press (1967).

Smith, Norman Kemp. *The Philosophy of David Hume.* London: Macmillan (1969).

Tononi. Phi: *A Voyage from the Brain to the Soul* (2012).

Peter Woit. *Not Even Wrong: The Failure of String Theory and the Continuing Challenge to Unify the Laws of Physics.*

Chapter 1

1. Mora, C., Tittensor, D. P., Myers, S. A., Worm, B., Davies, P., Ehrlich, S., Lovejoy, T. E. (2011). *How many species are there on Earth and in the ocean?* A massively distributed estimation approach. *Nature*, 486(7400), 478–486. https://journals.plos.org/plosbiology/article?id=10.1371/journal.pbio.1001127

2. IUCN *Red List of Threatened Species*. Version 2021-3. https://www.iucnredlist.org/

3. Wilson, E. O. (1992). *The diversity of life*. W.H. Freeman and Company.

4. May, R. M. (2000). *The abundance, diversity and extinction of tropical rain forest species*. Philosophical Transactions of the Royal Society B: Biological Sciences, 355(1403), 881–893.

Further reading

Berkeley, G. (1948–1957). *The Works of George Berkeley, Bishop of Cloyne*. A.A. Luce and T.E. Jessop (eds.). London: Thomas Nelson and Sons. 9 vols.

May, R. M. (2000). The abundance, diversity and extinction of tropical rain forest species. Philosophical Transactions of the Royal Society B: Biological Sciences, 355(1403), 881–893. https://doi.org/10.1098/rstb.2000.0714: https://doi.org/10.1098/rstb.2000.0714

Mora, C., Tittensor, D. P., Myers, S. A., Worm, B., Davies, P., Ehrlich, S., ... and Lovejoy, T. E. (2011). How many species are there on Earth and in the ocean? A massively distributed estimation approach. Nature, 486(7400), 478–486.

Wilson, E. O. (1992). *The diversity of life*. W.H. Freeman and Company.

Chapter 2

1. Blackmore, Susan. *Consciousness: An Introduction* (2nd ed., 2010).
2. Darwin, Charles. *The Descent of Man* (1871).
3. Darwin, Charles. *Autobiography* (1887).
4. Dawkins, Richard. *The Selfish Gene* (1976).
5. Dawkins, Richard. *The God Delusion* (2006).
6. Chalmers, D. J. (1995). *The conscious mind: In search of a fundamental theory*. New York: Oxford University Press.
7. Chalmers, D. J. (2002). The hard problem of consciousness. In D. J. Chalmers (Ed.), *Philosophy of mind: Classical and contemporary readings* (pp. 200–211). New York: Oxford University Press.
8. Chalmers, D. J. (2010). *The character of consciousness*. New York: Oxford University Press.
9. Searle, J. R. (1992). *The rediscovery of the mind*. Cambridge, MA: MIT Press.
10. Searle, J. R. (1997). *The mystery of consciousness*. New York: New York Review Books.
11. Searle, J. R. (2004). *Mind: A brief introduction*. New York: Oxford University Press.
12. Dennett, D. C. (1991). *Consciousness explained*. Boston: Little, Brown.
13. Dennett, D. C. (2006). *Sweet dreams: Philosophical investigations of sleep, dreaming, and the nature of consciousness*. Cambridge, MA: MIT Press.
14. Dennett, D. C. (2017). *From bacteria to Bach: The evolution of minds*. New York: W. W. Norton & Company.

15. James, William. *The Principles of Psychology* (1890).

16. Tononi, G. (2008). Consciousness as integrated information: A provisional theory. *BMC Neuroscience*, 9(5), 32.

17. Tononi, G. (2012). Integrated information theory: An overview. *Progress in Brain Research*, 193, 139–156.

18. Tononi, G. (2018). *The neural correlates of consciousness: An update on integrated information theory*. In B. Baumard and F. Metzinger (Eds.), *Conscious states: Computation, phenomenal experience, and the binding problem* (pp. 130–149). Oxford University Press.

19. Tononi, G. (2012). *Integrated information theory: From consciousness to coma*. Springer.

20. Tegmark, M. (2014). *Life 3.0: Being human in the age of artificial intelligence*. New York: Knopf.

21. Tegmark, M. (2016). *The mathematical universe: My quest for the ultimate nature of reality*. New York: Knopf.

22. Tegmark, M. (2020). *Our mathematical universe: My quest for the ultimate nature of reality*. New York: Knopf.

23. Tononi, G., and Koch, C. (2015). *Consciousness*.

24. Dorkenwald S, Matsliah A, Sterling AR, Schlegel P, Yu SC, McKellar CE, Lin A, Costa M, Eichler K, Yin Y, Silversmith W, Schneider-Mizell C, Jordan CS, Brittain D, Halageri A, Kuehner K, Ogedengbe O, Morey R, Gager J, Kruk K, Perlman E, Yang R, Deutsch D, Bland D, Sorek M, Lu R, Macrina T, Lee K, Bae JA, Mu S, Nehoran B, Mitchell E, Popovych S, Wu J, Jia Z, Castro M, Kemnitz N, Ih D, Bates AS, Eckstein N, Funke J, Collman F, Bock DD, Jefferis GSXE, Seung HS, Murthy M. *Neuronal wiring diagram of an adult brain.* FlyWire Consortium. bioRxiv [Preprint]. 2023 Jul 11:2023.06.27.546656. doi: 10.1101/2023.06.27.546656. PMID: 37425937; PMCID: PMC10327113.

Further reading

Baars, B. J. (1998). *A cognitive theory of consciousness.* Cambridge, MA: MIT Press.

Graziano, M. S. A. (2011). *Consciousness and the social brain.* Oxford, UK: Oxford University Press.

IUCN Red List: https://www.iucnredlist.org/

Malakar, R. (2016). Information conservation theory of consciousness: A new perspective on the nature of reality. *Frontiers in Psychology*, 7, 1696.

Malakar, R. (2017). Information conservation theory of consciousness: A computational model. *Frontiers in Psychology*, 8, 1728.

Mancuso, S., and Viola, G. (2018). *The conscious plant: A scientific exploration into the green kingdom.* MIT Press.

Mancuso, S., and Viola, G. (2019). *The quantum plant: A new scientific theory of consciousness.* MIT Press.

Mora et al. study: https://www.pnas.org/doi/full/10.1073/pnas.162359199

Penrose, R. (2005). *Cycles of time: An extraordinary new view of the universe.* London: Jonathan Cape.

Penrose, R. (2010). *The road to reality: A complete guide to the laws of the universe.* New York: Vintage Books.

Penrose, R. (2020). *Fashion, faith and fantasy in the new physics of the universe*. London: Vintage Books.

Tononi, G. (2012). *Integrated information theory: From consciousness to coma*. Springer.

Tononi, G., and Koch, C. (2015). *Consciousness: Towards a science of the subjective*. Springer.

Chapter 3

1. Darwin, Charles. *On the Origin of Species* (1859).
2. Shannon, Claude. *A Mathematical Theory of Communication* (1948).
3. David Deutsch. *The Fabric of Reality* (1997).
4. David Deutsch. *The beginning of infinity* (2011).
5. https://youtube.com/shorts/8r-ZQO74olY?si=cyxHeJ_FwyLWhJo1
6. Anil Seth. *Trends in Cognitive Sciences* (2013).
7. Anil Seth. *A predictive processing theory of sensorimotor contingencies* (2014).
8. Anil Seth. *The cybernetic Bayesian brain* (2015).

Further reading

Adami, Christoph. Information-theoretic considerations concerning the origin of life. *Origins of Life and Evolution of the Biosphere* 25.4 (1995).

Dawkins, Richard. *The Code of Codes: The Language of Life and How It Works* (1999).

Eigen, Manfred and Winkler-Oswatitsch, Ruthild (1983). *The Evolution of the Genetic Code.*

Frankham, Richard, Eldridge, Michael D., and Crawford, David I. (2023). *Genetics and the Conservation of Evolving Populations.*

Klug, William S. and Cummings, Michael R. (2017). T*he Universal Genetic Code.*

Meffe, Gary K. and Carroll, C. R. (2014). *Principles of Conservation Biology.*

Nirenberg, Marshall (1966). *The Genetic Code: The Universal Language of Life.*

Orgel, Leslie (1973). T*he Origin of Life: A Theory of Genetic Information.*

Penrose, R. (2005). *Cycles of time: An extraordinary new view of the universe.* London: Jonathan Cape.

Penrose, R. (2010). *The road to reality: A complete guide to the laws of the universe.* New York: Vintage Books.

Penrose, R. (2020). F*ashion, faith and fantasy in the new physics of the universe.* London: Vintage Books.

Schaal, Bradley C., Schaal, Bruce M., and Le Corff, William J. (2008). *Conservation Genetics.*

Schwartz, Mark, Tallmon, Daniel N., and Luikart, Greg (2010). *The Conservation Genetics Handbook.*

Shannon, Claude E. A mathematical theory of communication. *Bell system technical journal* 27.3 (1948): 379–423.

Stephen H. Schneider and David L. Semenov (2016). *Conservation Biology: Foundations, Concepts, Applications.*

Chapter 4

1. Penrose, Roger. *The Emperor's New Mind* (1989).
2. Penrose, Roger. *Shadows of Mind* (1994).
3. Arkani-Hamed, Nima. An interview with YouTube channel Closer to Truth.
4. Susskind, Leonard (2008). The Black Hole War: My Battle with Stephen Hawking to Make the World Safe for Quantum Mechanics.
5. David Deutsch YouTube Channel – Constructor Theory Lecture.

Further reading

Al-Khalaili, J., and McFadden, J. (2014). *Life on the edge: the coming of age of quantum biology.*

Arkani-Hamed, Nima: *an interview with YouTube channel Closer to Truth.*

Atwal, G. S., Hameroff, S. R., Tuszynski, J. A., and Koonin, E. V. (2000). *The cyton skeleton as a possible substrate for quantum computation in the brain.* Trends in Cognitive Sciences, 4(2), 41–49.

Barger, V., and Marciano, W. (2010). *The anthropic principle and the fine-tuning of the universe.* Annual Review of Nuclear and Particle Science, 60(1), 537–566.

Nick Bostrom, Nick (2002). *Must the Universe Spawn Life and Mind?*

Chalmers, David Chalmers (1995). *The Hard Problem of Consciousness.*

Craddock, Travis JA, Tuszynski, Jack A, Hammersoff, Stuart. (2012). Cytoskeletal Signaling: *Is Memory Encoded in Microtubule Lattices by CaMKII Phosphorylation?* PLOS Computational Biology.

Davies, P. (2007). Cosmic coincidences: *Dark matter, dark energy, anthropic design.* New York: Houghton Mifflin Harcourt.

Demuth, H. and M. Beale, M. (2000). *Neural Network Toolbox: for use with* MATLAB, The MATH WORKS Inc. (2000).

Dennett, Daniel (1991). *Consciousness Explained* (1991).

Deutsch, David (1999). *The Beginning of Infinity: Explanations That Transform the Way We Think About the Universe* (1999).

Deutsch, David. *Constructor Theory Lecture,* David YouTube Channel.

Doya, K. *What are the computations of the cerebellum, the basal ganglia, and the cerebral cortex,* Neural Networks, Vol. 12, pp. 961–974 (1999).

Drake, Frank (1961). *The Green Bank Conference* (1961).

Goodfellow, I., Yoshua, B., Courville, A. *Deep learning*. MIT Press (2016).

Hagan M., Menhaj, M. *Training feedforward networks with the Marqurdt algorithm,* IEEE Trans. on Neural Networks, Vol. 5, No. 6, pp. 989-993 (1994).

Hagan, M. T., Demuth, H. B. and M. H. Beale (1996). *Neural Network Design,* PWS Publishing, Boston (1996).

Hameroff, S. R. (2013). *Consciousness in the universe: A review of the 'Orch OR' theory.* Physics of Life Reviews, 11(1), 39–78.

Hameroff, S. R. (2014). *Quantum walks in brain microtubules—A biomolecular basis for quantum cognition?* Topics in Cognitive Science, 6(1), 91–97.

Hameroff, Stuart and Penrose, Roger. (1996). *Conscious events as orchestrated spacetime selections.* Journal of Consciousness Studies.

Hebb, D. O. (1949). *The Organization of Behavior - A Neuropsychological Theory,* John Wiley & Son, New York (1949)

Iwata, A., K. Hotta, K., H. Matsuo, H. and N. Suzumura, N. *Large scale neural network "CombNET",* IEICE Trans. Inf. and Syst, Vol. J73-D-II, No. 8, pp. 1261–1267 (1990–8) (in Japanese).

Iwata, Akira, Kenichi Hotta, Kenichi, Keishi Matsuo, and Suzumura, Nobuo. *Large-scale 4-layer neural network "CombNET",* IEICE, J73-D-II, 8, pp. 1261–1267 (1990–8).

James, Gareth, Daniela Witten, Trevor Hastie, and Robert

Tibshirani. *An introduction to statistical learning: with applications. R. Springer Science & Business Media* (2013).

Kohonen, T. *Self-Organizing Maps*, 3ed, Springer, Heidelberg.

Levenberg, K. (1944). *A method for the solution of certain problems in least squares*, Quart. Appl. Math, Vol. 2, pp. 164–168.

Lillicrap, T. P., Hunt, J. J., Pritzel, A., Heess, N., Erez, T., Tassa, Y., ... and Silver, D. (2015). *Human-level control through deep reinforcement learning*. Nature, 529(7585), 464–468.

Marquardt, D. *An algorithm for least-squares estimation of nonlinear parameters*, SIAM J. Appl. Math, Vol. 11, pp. 431–441 (1963).

Mnih, V., Kavukcuoglu, K., Silver, D., Graves, A., Antonoglou, I., Wierstra, D., and Riedmiller, M. (2013). *Playing Atari with deep reinforcement learning*. arXiv preprint arXiv:1312.5602.

Mnih, V., Kavukcuoglu, K., Silver, D., Graves, A., Antonoglou, I., Wierstra, D., and Riedmiller, M. (2015). *Human-level control through deep reinforcement learning*. Nature, 518(7540), 529–533.

Murphy, Kevin P. Machine learning: A probabilistic perspective. MIT Press (2012).

Paraskecopoulos, V., Heywood, M. I. and Chatwin, C. R. Modular SRV *Reinforcement learning architectures for non-linear control,* International Journal of Intelligent Control and Syst., Vol. 3(2), pp. 171–192 (1999).

Penrose, R., and Hameroff, S. R. (1995). *Shadows of the mind:*

A search for the missing science of consciousness. Oxford University Press.

Penrose, R., and Hameroff, S. R. (2014). *Reply to criticism of the 'Orch OR' qubit–'Orchestrated objective reduction' is scientifically justified.* Physics of Life Reviews, 11(11), 1877–1890.

Penrose, Roger (2006). *Before the big bang: An outrageous new perspective and its implications for particle physics* (2006).

Penrose, Roger Penrose. *Cycles of time: An extraordinary new view of the universe* (2010).

Piccinini, Gualtiero Piccinini and Chalmers, David. *Panpsychism: The View That Everything Has a Mind* (2002).

Sahu, S., Ghosh, S., Ghosh, B., Aswani, K., Hirata, K., Fujita, D., and Bandyopadhyay, A. *Atomic water channel controlling remarkable properties of a single brain microtubule: Correlating single protein to its supramolecular assembly.* Biosensors & Bioelectronics (2013).

Sakakawa, T., Hu, J. and Hirasawa, K. *Self-organizing function localization neural network*, Trans. of the Society of Instrument and Control Engineers, Vol. 41, No. 1, pp. 67–74 (2005–1) (in Japanese).

Sawaguch, Toshiyuki. *Brain structure and evolution of intelligence*, Kaimeisha (1989).

Schulman, J., Levine, S., Abbeel, P., Jordan, M., and Moritz, P. (2015). Trust region policy optimization. arXiv preprint arXiv:1502.05477.

Solla, S. A. and M. Fleisher, M. *Generalization in feedforward neural networks, in Proc.* The IEEE International Joint Conference on Neural Networks (Seattle), pp. 77–82 (1991).

Susskind, L. *The anthropic principle.* arXiv preprint arXiv:1308.0787 (2013).

Sutton, R. S. and Barto, A. G. *Reinforcement Learning: An Introduction,* MIT Press, Cambridge (1998).

Sutton, R. S., and Barto, A. G. *Reinforcement learning: An introduction.* MIT Press (2018).

Tegmark, Max. *Importance of quantum decoherence in brain processes.* Physical Review (2000).

Tesauro, G. J. TD-Gammon, *A self-teaching backgammon program, achieves master-level play,* Neural Computation, Vol. 6, pp. 215–219 (1994).

Turing, Alan. *Computing Machinery and Intelligence.* (1950).

Wallace, Alan. *The Conscious Universe: The Scientific and Spiritual Implications of a Panpsychist Worldview* (2011).

Chapter 5

1. Financial Times. Can man ever build a mind? 11/1/2019.
2. Penrose, Roger. The Emperor's New Mind (1989).
3. Penrose, Roger. Shadows of Mind (1994).

4. Hameroff, Penrose. Conscious events as orchestrated space-time selections (1996).
5. Hameroff, Penrose. Consciousness in the universe, a review 'Orch OR' Theory (2014).

Further reading

Damasio, A. R. *The strange order of things: What neuroscience can tell us about the origins of human emotions and the brain's modular design.* Penguin Books (2018).

Dennett, D. C. *Consciousness explained*, Little, Brown and Company (1991).

Domingos, P. *The Master Algorithm: How the Quest for the Ultimate Learning Machine Will Remake Our World*, Basic Books (2015).

Goodfellow, I., Bengio, Y., Courville, A., and Bengio, Y. Deep *learning* (Vol. 1). MIT press Cambridge (2016).

LeCun, Y., Bengio, Y., and Hinton, G. *Deep learning*. Nature, 521(7553), 436–444 (2015).

McCulloch, W. S., and Pitts, W. *A logical calculus of the ideas immanent in nervous activity*. Bulletin of Mathematical Biophysics, 5(4), 115–133 (1943).

Pinker, S. *The blank slate: The modern denial of human nature.* Viking (2007).

Rosenblatt, F. *The perceptron: A probabilistic model for information storage and organization in the brain.* Psychological Review, 65(6),

386–408 (1958).

Samuel, A. L. *Some studies in machine learning using the game of checkers.* IBM Journal of Research and Development, 3(3), 210–229 (1959).

Schmidhuber, J. *Deep learning in neural networks: An overview.* Neural Networks, 61, 85–117 (2015).

Sutton, R. S., and Barto, A. G. *Reinforcement Learning: An Introduction*, MIT Press (2018).

Taylor, C. *A secular age.* Harvard University Press (2007).

Wolf, E. B. *In praise of shadows* (2010).

Chapter 6

1. McCulloch, W. S., and Pitts, W. A logical calculus of the ideas immanent in nervous activity. Bulletin of Mathematical Biophysics, 5(4), 115–133 (1943).
2. Rosenblatt, F. The perceptron: A probabilistic model for information storage and organization in the brain. Psychological Review, 65(6), 386–40 (1958).
3. Hinton, G. E., Srivastava, N., Krizhevsky, A., Sutskever, I., and Salakhutdinov, R. R. *Improving neural networks by preventing co-adaptation of feature detectors.* arXiv preprint arXiv:1207.0580 (2012).
4. LeCun, Y., Bottou, L., Bengio, Y., and Haffner, P. *Gradient-based learning applied to document recognition.* Proceedings of the IEEE, 86(11), 2278–2324 (1998).
5. Donald Hebb. *The Organisation of Behaviour* (1949).

6. George Cybenko. *Approximation by Superpositions of a Sigmoidal Function*, Mathematics of Control, Signals, and Systems (MCSS), Volume 2 (1989).

Chapter 7

1. The Independent. *Google fires software engineer who claimed its AI had become sentient and self-aware*: https://www.independent.co.uk/tech/google-ai-sentient-self-aware-blake-lemoine-b2130634.html

2. https://twitter.com/60Minutes/status/1647742247444553732?lang=en

Chapter 8

1. Dennett, D. C. *Consciousness explained*. Boston: Little, Brown (1991).
2. Friston, K. J., Parr, T., and Stephan, K. E. *Active inference and counterfactual thinking: A unification*. Neural Computation, 34(1), 144–187 (2022).
3. Fristo, K.J. Kilner, J. and Harrison, L. *The free-energy principle: a unified brain theory?* Nature Reviews Neuroscience 2006.
4. Daniel Kahneman and Amos Tversky. Heuristics and Biases, The Psychology of Intuitive Judgment (2002).
5. Wolfram, S. *A New Kind of Science* (2nd ed.). Wolfram Media (2012).
6. Wolfram, S. *The Concept of the Ruliad*. Wolfram Research (2021).

7. Friston, K. J., and Stephan, K. E. *Counterfactual inference and the free energy principle.* In The Oxford Handbook of Cognitive Psychology (pp. 451–474). Oxford University Press (2022).

Further reading

Friston, K. J. *Free-energy minimization and the dark-room problem.* Journal of Physiology-Paris 2009.

Friston, K. J. *The free energy principle and counterfactuals: Linking perception, action, and decision-making.* Frontiers in Psychology, 14, 713955 (2023).

Friston, K.J. Harrison, L. and Penny, W., *A theory of cortical responses, Published in:* Philosophical Transactions of the Royal Society B: Biological Sciences (2003).

Friston, K. J., Parr, T., and Stephan, K. E. *Active inference and counterfactual thinking: A unification.* Neural Computation, 34(1), 144–187 (2022).

Friston, K. J., and Stephan, K. E. *Counterfactual inference and the free energy principle.* In The Oxford Handbook of Cognitive Psychology (pp. 451–474), Oxford University Press (2022).

Friston, K. J. *The free energy principle and counterfactuals: Linking perception, action, and decision-making.* Frontiers in Psychology, 14, 713955 (2023).

Gibbons, G. W., and Shellard, E. P. *The Temporal Topology of the Universe* (2013).

Gurzadyan, V., and Penrose, R. *Concentric circles in WMAP data may provide evidence of violent pre-Big-Bang activity* (2010).

Penrose, R. *Cycles of Time: An Extraordinary New View of the Universe* (2010).

Penrose, R. Fashion, Faith, and Fantasy in the New Physics of the Universe (2016).

Daniel Kahneman and Amos Tversky. Heuristics and Biases: *The Psychology of Intuitive Judgment* (2002).

Chapter 9

1. GONDRAN Alexandre, CC BY-SA 4.0 <https://creativecommons.org/licenses/by-sa/4.0>, via Wikimedia Commons
   https://commons.wikimedia.org/wiki/Commons:Creative_Commons_Attribution-ShareAlike_3.0_Unported_License
   https://commons.wikimedia.org/wiki/Commons:Licensing
   https://commons.wikimedia.org/wiki/File:Double-slit_experiment_with_electrons.png
   https://upload.wikimedia.org/wikipedia/commons/a/a2/Double-slit_experiment_with_electrons.png
2. David J. Griffiths (and Darrell F. Schroeter in later editions), *Introduction to Quantum Mechanics* (2018).
3. Sabine Hossenfelder, *Existential Physics* (2022).
4. Sabine Hossenfelder, *Lost in Math* (2018).
5. Wigner's paper, *Remarks on the Mind-Body Question, in the book The Scientist Speculates* (I.J. Good, ed.).

Further readings

Linde, A.D. P*article Physics and Inflationary Cosmology.* Harwood Academic Publishers (1990).

Chapter 10

1. David Deutsch: *The beginning of infinity* (2011).

2. https://www.youtube.com/watch?v=XZyLQr6kv3I&t=1736s

Further readings:

Don S Lemons: *A Student's Guide to Entropy* (2013).

Linde, A.D. *Particle Physics and Inflationary Cosmology,* Harwood Academic Publishers (1990).

Chapter 11

1. Wolfram, S. *A New Kind of Science* (2nd ed.). Wolfram Media. (2012).
2. Henri Bergson *Time and Free Will: An Essay on the Immediate Data of Consciousness* (1889): Bergson's doctoral dissertation.

Chapter 12

1. Friston, K. J., Parr, T., and Stephan, K. E. *Active inference and counterfactual thinking: A unification.* Neural Computation, 34(1), 144-187 (2022).

Further readings:

Albert, D. Z. *Time and Chance*, Harvard University Press. (2000).

Bergson, *Time and Free Will: An Essay on the Immediate Data of Consciousness* (1889).

Bergson, *Matter and Memory* (1896).

Bergson, *Creative Evolution* (1907).

Carroll, Sean. *Spacetime and Geometry: An Introduction to General Relativity*, W. H. Freeman & Company (2004).

Carroll, S. M. *From Eternity to Here: The Quest for the Ultimate Theory of Tim* (2010).

Dehaene S. *Consciousness and the Brain: Deciphering How the Brain Codes Our Thoughts.*

Einstein, Albert. *Relativity: The Special and General Theory*, Routledge (2016).

Greene, Brian. *Fabric of the Cosmos: Space, Time, and the Texture of Reality*. Penguin Books (2004).

Isaacson W. Einstein: *His Life and Universe.*

Koch, C., and Tononi, G. *Consciousness: Confidently unconscious,* MIT Press (2015).

Pearson K. A. and Mullarkey J. *Introduction to Bergson's philosophy:*

Henri Bergson: Key Writings.

Schurger Aaron, et al. *The Neural Basis of Free Will: Criterial Causation*

Symmetry in quantum Symmetry in quantum mechanics, https://en.wikipedia.org/wiki/Symmetry_in_quantum_mechanics

Is a time symmetric interpretation of quantum theory possible without retrocausality? https://royalsocietypublishing.org/doi/10.1098/rspa.2016.0607

A time-symmetric formulation of quantum mechanics, https://pubs.aip.org/physicstoday/article/63/11/27/413341/A-time-symmetric-formulation-of-quantum-mechanics

Time-symmetry https://en.wikipedia.org/wiki/T-symmetry

Tononi, G. *Consciousness: An integrated information theory,* Oxford University Press (2004).

Tononi, G., and Koch, C. *Consciousness: Does it matter?* Scientific American, 313(5), 38–45 (2015).

Zakay, J. *Time and Human Cognition: A Life-Span Perspective.*

Zeh, H. D. *The Physical Basis of the Direction of Time.* Springer (2007).